Memoirs of the American Mathematical Society
Number 343

Michael Aschbacher

Overgroups of Sylow subgroups in sporadic groups

Published by the
AMERICAN MATHEMATICAL SOCIETY
Providence, Rhode Island, USA

March 1986 · Volume 60 · Number 343 (end of volume)

MEMOIRS of the American Mathematical Society

SUBMISSION. This journal is designed particularly for long research papers (and groups of cognate papers) in pure and applied mathematics. The papers, in general, are longer than those in the TRANSACTIONS of the American Mathematical Society, with which it shares an editorial committee. Mathematical papers intended for publication in the Memoirs should be addressed to one of the editors:

PREPARATION OF COPY. Memoirs are printed by photo-offset from camera-ready copy prepared by the authors. Prospective authors are encouraged to request a booklet giving detailed instructions regarding reproduction copy. Write to Editorial Office, American Mathematical Society, Box 6248, Providence, RI 02940. For general instructions, see last page of Memoir.

SUBSCRIPTION INFORMATION. The 1986 subscription begins with Number 339 and consists of six mailings, each containing one or more numbers. Subscription prices for 1986 are $214 list, $171 institutional member. A late charge of 10% of the subscription price will be imposed on orders received from nonmembers after January 1 of the subscription year. Subscribers outside the United States and India must pay a postage surcharge of $18; subscribers in India must pay a postage surcharge of $15. Each number may be ordered separately; *please specify number* when ordering an individual number. For prices and titles of recently released numbers, see the New Publications sections of the NOTICES of the American Mathematical Society.

BACK NUMBER INFORMATION. For back issues see the AMS Catalogue of Publications.

Subscriptions and orders for publications of the American Mathematical Society should be addressed to American Mathematical Society, Box 1571, Annex Station, Providence, RI 02901-1571. *All orders must be accompanied by payment.* Other correspondence should be addressed to Box 6248, Providence, RI 02940.

MEMOIRS of the American Mathematical Society (ISSN 0065-9266) is published bimonthly (each volume consisting usually of more than one number) by the American Mathematical Society at 201 Charles Street, Providence, Rhode Island 02904. Second Class postage paid at Providence, Rhode Island 02940. Postmaster: Send address changes to Memoirs of the American Mathematical Society, American Mathematical Society, Box 6248, Providence, RI 02940.

Printed in the United States of America.

The paper used in this journal is acid-free and falls within the guidelines established to ensure permanence and durability.

TABLE OF CONTENTS

ABSTRACT

The maximal overgroups of noncyclic Sylow subgroups of the sporadic finite simple groups are determined. Moreover a geometric structure is associated to this collection of overgroups, which is useful in the study of the sporadic groups.

1980 Mathematics Subject Classification: 20B25, 20D08, 20D05.

Library of Congress Cataloging-in-Publication Data

Aschbacher, Michael, 1944–
Overgroups of Sylow subgroups in sporadic groups.

(Memoirs of the American Mathematical Society, ISSN 0065-9266; no. 343 (March 1986)
"Volume 60."
Bibliography: p.
1. Sporadic groups (Mathematics) 2. Sylow subgroups. I. Title. II. Series.
Memoirs of the American Mathematical Society; no. 343.
QA3.A57 no. 343 [QA171] 510s [512'.2] 86-1040
ISBN 0-8218-2344-2

1. INTRODUCTION.

Let G be a sporadic finite simple group and p a prime such that p^2 divides the order of G. In this paper the maximal subgroups of G containing a Sylow p-subgroup T of G are essentially determined up to conjugacy. Moreover a geometric structure is associated to the collection of overgroups of T which is useful in the study of the subgroup structure and the permutation representations of G.

The notion of geometry used here was developed many years ago by J. Tits to study algebraic groups. More recently F. Buekenhout observed that the theory could be extended to include many geometries associated to sporadic finite simple groups. [45] To my knowledge Ronan and Smith were the first to concentrate on geometries associated to p-locals in sporadic groups. [46] Definitions of geometric terminology are collected in section 27. In particular we find that given a group G and a family $\mathcal{F}$ of subgroups of G there exists a coset geometry $\Gamma(G,\mathcal{F})$ upon which G is represented as a group of automorphisms.

When the p-part of the order of the sporadic group G is p^2, the existence of subgroups of G isomorphic to $L_2(p^2)$ is sometimes left open in this paper, although such subgroups are sometimes known to exist from other sources. To be specific, this is done when p is 3,5,5,7 and G is HS, Sz, M(23),F_2, respectively. Similarly when p = 3 and G is HS the existence of M_{11} and $L_3(4)$ subgroups is left open and when p is 3 and G is J_4 the existence of maximal $U_3(3)$ and $PGL_3(4)$ subgroups is left open. (Of course all the maximal subgroups of HS are known already by Magliveras.) The emphasis here is on large maximal subgroups. The cases left open are in some sense too small to be treated elegantly using the methods of this paper.

Let G be a finite group, p a prime, and T a Sylow p-subgroup of G. Define a *p-basis* for G to be a family $\mathcal{F}$ = $(G_i : i \in I)$ such that $\mathcal{F}$ is a set of representatives for the $N_G(T)$-orbits of maximal overgroups of T, such that the

---

Received by the editors February 5, 1985.

Partial support supplied by the National Science Foundation.

map $J \mapsto G_J$ is a bijection of the power set of I with the orbits of $N_G(G_I)$ on overgroups of G_I in G, and such that $N_G(G_J)$ is transitive on the G-conjugates of G_K in G_J for all subsets J and K of I. The basis is *well behaved* if $N_G(T) = N_{G_i}(T)N_{G_j}(T)$ for all distinct i and j in I and G is flag transitive on $\Gamma(G,\mathcal{F})$.

We find in 28.2 that if G possesses a well behaved p-basis then every set of representatives for the $N_G(T)$-orbits of maximal overgroups of T is a p-basis and that $N_G(T)$ is transitive on the bases over T. By 28.1, $\Gamma(G,\mathcal{F})$ is residually connected. The overgroups of conjugates of G_I will be termed *parabolics*. By 28.3 each overgroup Y of T in G determines uniquely up to conjugation in $N_G(T)$ a smallest parabolic over Y. Thus all overgroups of T in G are described via the basis, as we also know the overgroups of T in each parabolic.

Of course if G is a finite group of Lie type and characteristic p then G possesses a well behaved p-basis and the parabolics of G are just the usual parabolics. This is important since one of the objects of this exercise is to produce descriptions of subgroup structure applicable to as wide a class of groups as possible. More on that later after the statement of the following result:

Theorem 1. Let G be a sporadic finite simple group and p a prime such that p^3 divides the order of G. Then either G possesses a well behaved p-basis or one of the following holds:

(1) p = 2 and G is M_{11}, M_{23}, Mc, He, or M(23).

(2) p = 3 and a Sylow 3-subgroup of G is isomorphic to 3^{1+2}.

(3) p = 5 and G is Ru or F_3.

(4) p = 13 and G is F_1.

Associated to each group G and p-basis $\mathcal{F}$ is a diagram. For example if G is of Lie type and characteristic p, that diagram is just the Dynkin diagram. The diagram provides a compact means of describing portions of the structure of the groups $(G_J : J \subseteq I)$ and the geometry $\Gamma(G,\mathcal{F})$. The diagrams for the sporadic groups and tables describing the maximal overgroups of Sylow groups appear at the end of this section. The notational conventions defining the diagrams are expained in section 27. Many of these conventions come from Ronan-Smith in [46] or Buekenhout in [45]. Many of the geometries listed are already known.

In a group G of Lie type in characteristic p, the Borel-Tits Theorem provides a description of the p-local structure of G in terms of the parabolics. In a later paper an analogue of the Borel-Tits Theorem will be established describing the p-local structure of each sporadic group in terms of its p-parabolics.

The following results are established in the course of determining overgroups of Sylow groups and may be of independent interest:

Theorem 2. Let G be a finite group and A a noncyclic elementary abelian p-subgroup of G such that A is contained in a unique maximal subgroup M of G. Assume M contains no nontrivial normal subgroup of G. Then one of the following holds:

(1) G is a group of Lie type in characteristic p and Lie rank 1 and M is a Borel subgroup of G.

(2) G is an alternating group of degree 2p, $A \cong E_{p^2}$, and M is of index 2 in $A_p \text{ wr} Z_2$.

(3) G is Aut(Sz(32)), $A \cong E_{25}$, and M is a Sylow 5 normalizer.

Theorem 3. Let G be a group with $F^*(G) = L \cong PSL(V)$, PSp(V), or PSU(V), where V is an n-dimensional vector space over a finite field of order q.

(1) If U is a proper nondegenerate subspace of V with $\dim(U) \neq \dim(V/U)$ then $N_G(U)$ is a maximal subgroup of G.

(2) Let V be the direct sum of a set Σ of $m > 1$ subspaces of dimension $k > 1$ and if L is symplectic or unitary assume the sum is orthogonal. Then either $N_G(\Sigma)$ is maximal in G or each of the following hold:

(a) $k = 2$.

(b) $q = 2$ or L is unitary and $q = 4$.

(c) $N_G(\Sigma) \leq N_G(H)$ where one of the following holds:

(i) $H \cong Sp_n(2)$ and $L \cong L_n(2)$.

(ii) $H \cong O_n^{\epsilon}(2)$ with $\epsilon = (-1)^m$ and $L \cong Sp_n(2)$.

(iii) $H \cong S_n/E_{3^{n-1}}$ or $U_4(3)$ and $L \cong U_n(2)$, with $n = 6$ if H is $U_4(3)$.

In section 7 the analogue of Theorem 3 is established when L is an alternating group.

The proofs of these results and the determination of overgroups involves an appeal to the classification of the finite simple groups. Alternatively the results can be regarded as theorems about K-groups. The latter is a useful point of view as much of the information generated here will presumably be useful in future simplifictions and revisions of the proof of the Classification Theorem.

In investigating the sporadic groups I have chosen to give myself the order and the general structure of the normalizers of subgroups of prime order in each group as a point of departure. A list of these properties appears in section 5 of [23]. The subgroup structure of the Mathieu groups seems best studied from the point of view of the representation of M_{24} on its Steiner system. Thus I also give myself the table of maximal subgroups of the Mathieu groups appearing in [19]. Effectively this means I prove nothing about the Mathieu groups. Also the representation of .0 on the Leech lattice is used to infer the existence of various containments among subgroups of Co_1. To establish the existence of a $U_3(11)$ subgroup of J_4, it is assumed that J_4 possesses a faithful 112 dimensional module over the field of order 2. Some of Parrott's work on Ru and F_3 in [32] and [33] is assumed.

Certainly these are not ideal assumptions with which to begin. A selfcontained development of the subgroup structure of the sporadic groups would presumably start with the definition of each group either as a stabilizer in the group of automorphism of some convenient structure (Such as the Steiner system for Mathieu groups or the Leech lattice for certain subgroups of Co_1.) or as a group possessing some particular centralizer. The structure of the normalizers of subgroups of prime order and the order of the sporadic group would presumably be derived more or less simultaneously with that of some of the overgroups.

A major objective has been to find points of view which allow us to make and prove statements about classes of simple groups rather than to study each sporadic group individually. The most successful example of this involves the notion of groups of GF(p)-type. A group G is of *GF(p)-type* if it possesses a *large symplectic p-subgroup* Q. That is Q satisfies the following:

(L1) Q is a p-subgroup of G of symplectic type and exponent p
or 4.

(L2) $C_G(Q) \leq Q$.

(L3) $Q \trianglelefteq N_G(\Omega_1(Z(Q)))$.

(L4) $\mathcal{U}_G(Q,p') = 1$.

Most sporadic groups which are divisible by a high power of p are of

GF(p)-type.

Of course there already exist in the literature lists of maximal subgroups of many of the sporadic groups. For example Robert Wilson has recently provided such lists for several more of the sporadic groups. My objectives are somewhat different. In addition to enumerating subgroups, I'm interested in providing points of view and systematizations which improve our understanding of subgroup structure and facilitate the proof of group theoretic results.

The tables at the end of this section utilize the following conventions: Let P be a maximal parabolic and $R = O_2(P)$. The general structure of P/R is listed in column 2. $H_1/H_2/\ldots$ denotes a group K with a normal series $K = K_0 > K_1 > \ldots$ such that $K_{i-1}/K_i \cong H_i$. The p-rank of the p-chief factors of P are listed in column 3; $n_1/n_2/\ldots$ indicates the existence of a P-chief series $R = R_0 > R_1 > \ldots$ with $m(R_{i-1}/R_i) = n_i$. Column 4 records the number of conjugates of P containing a Sylow p-subgroup T of G. Thus the entry in column 4 is $|N_G(T):N_P(T)|$.

Column 5 contains various remarks about the parabolic. If G is of GF(p)-type then Q denotes its large symplectic subgroup and H is $N_G(Q)$. When G is of GF(p)-type the isomorphism type $Q \cong p^{1+2n}$ or $Q \cong D^n$, D^nQ, or Z_4*D^n will be listed. Sometimes column 5 will contain the P-somorphism type of the P-series of R described in column 3. $V_1 \otimes V_2$ denotes a module for $H_1 \times H_2$ which is the tensor product of the modules V_i for H_i. $L_n(p)^{(i)}$ denotes the ith fundamental irreducible module for $L_n(p)$. Thus for example $L_n(p)^{(i)}$ is of dimension $\binom{n}{i}$, $L_n(p)^{(1)}$ is the natural module, and $L_n(p)^{(n-1)}$ is its dual. $\Omega_n^\epsilon(p)$ denotes the n-dimensional orthogonal module for $\Omega_n^\epsilon(p)$. A_6 denotes either of the 4 dimensional GF(2)-modules for A_6. A_6/Z_3, M_{22}/Z_3, and $U_4(3)/Z_3$ denote the irreducible GF(4)-modules of dimension 3, 6, and 6, for the respective groups. Todd and dual Todd modules for the Mathieu groups are discussed in section 31.

When appropriate, a diagram is listed for the group. The diagrams are discussed in section 27.

At the end of the collection of tables the p-parabolics of certain groups of Lie type over GF(r) are listed. In each case (p,r) = (2,3) or (3,2). These groups are all of GF(p)-type.

Most of the notation and terminology in this paper is standard; [6] is a reference for such things. This section concludes with the definitions of a few peices of specialized notation and terminology.

Given a group G and a subgroup X of G, $\mathcal{M} = \mathcal{M}_G$ denotes the set of maximal subgroups of G and $\mathcal{M}(X) = \mathcal{M}_G(X)$ denotes the set of those member of $\mathcal{M}$ containing X.

Let p be a prime. Chev(p) denotes the set of simple groups of Lie type and characteristic p. $G_2(2)' = U_3(3)$, $Sp_4(2)' = A_6$, and ${}^2F_4(2)'$ are regarded as members of Chev(2) and ${}^2G_2(3)' = L_2(8)$ is regarded as a member of Chev(3). For $G \in$ Chev(p), G is said to be *defined over the field* F if F is the order of a long root group of G. $L_n^+(q) = L_n(q)$, $L_n^-(q) = U_n(q)$, $E_6^+(q) = E_6(q)$, and $E_6^-(q) = {}^2E_6(q)$.

The author would like to thank Professor S. Smith for suggestions leading to numerous improvements in this manuscript.

M_{11}

$p = 2$; $|T| = 2^4$

1	M_{10}	0	1	Stabilizer of 1 of 11 points
2	SD_{16}/E_9	0	1	Global stabilizer of 2 of 11 points. Normalizer of Sylow 3-group.
3	S_3	2/1	1	Global stabilizer of 3 of 11 points. Centralizer of involution. $P \cong GL_2(3)$.

$p = 3$; $|T| = 9$

1	M_{10}	0	2	Stabilizer of 1 of 11 points.
2	SD_{16}/E_9	2	1	Global stabilizer of 2 of 11 points. Normalizer of Sylow 3-group.

M_{12}

$p = 2$; $|T| = 2^6$; $Q \cong D^2$

1	S_3	(2+2)/1	1	H. Both Q_8's are normal in H.
2	S_3	1/2/2	1	R is Z_4^2 inverted by involution.

$p = 3$; $|T| = 3^3$; $Q \cong 3^{1+2}$

1	$GL_2(3)$	2	1	The two maximal parabolics
2	$GL_2(3)$	2	1	are fused in Aut(G).

M_{22}

$p = 2$; $|T| = 2^7$; $Q \cong D^2$ 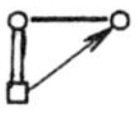

1	S_5	4	1	$L_2(4)$-module.
2	A_6	4	1	A_6-module.

$p = 3$; $|T| = 9$

1	$L_3(4)$	0	1	Stabilizer of 1 of 22 points.
2	A_6/E_{16}	0	2	
3	A_7	0	2	
4	M_{10}	0	1	

M_{23}

$p = 2$; $|T| = 2^7$

1	M_{22}	0	1	Stabilizer of 1 of 23 points.
2	$Z_2/L_3(4)$	0	1	Global stabilizer of 2 points.
3	A_7	4	1	Global stabilizer of septad.
4	$Z_2/GL_2(4)$	4	1	$L_2(4)$-module.

$p = 3$; $|T| = 9$

1	M_{22}	0	2	Stabilizer of 1 of 23 points.
2	$Z_2/L_3(4)$	0	2	Global stabilizer of 2 points.
3	A_7/E_{16}	0	4	Global stabilizer of septad.
4	$Z_2/GL_2(4)/E_{16}$	0	4	Global stabilizer of 3 points.
5	A_8	0	2	
6	M_{11}	0	1	

M_{24}

$p = 2$; $|T| = 2^{10}$; $Q \cong D^3$ 

1	$L_4(2)$	4	1	Split extension; $L_4(2)$; Global stabilizer of octad.
2	$S_3 \times L_3(2)$	6	1	$S_3 \otimes L_3(2)$. Stabilizer of trio.
3	S_6/Z_3	6	1	A_6/Z_3. Stabilizer of sextet.

$p = 3$; $|T| = 3^3$

1	$S_3/L_3(4)$	0	2	Global stabilizer of 3 points.
2	$S_6/Z_3/E_{64}$	0	1	Stabilizer of trio.
3	Z_2/M_{12}	0	1	Transitive on 24 points.

J_1

$p = 2$; $|T| = 8$

1	Z_3/Z_7	3	1	Normalizer of T.
2	A_5	1	7	Centralizer of involution.

J_2

$p = 2$; $|T| = 2^7$; $Q \cong QD$

1	A_5	4/1	1	H; $\Omega_4^-(2)$-module.
2	$Z_3 \times S_3$	4/2	1	Normalizer of J(T). R Sylow in $L_3(4)$.

$p = 3$; $|T| = 3^3$

1	$PGL_2(9)$	1	1	$F^*(P)$ quasisimple.
2	$U_3(3)$	0	1	

$p = 5$; $|T| = 25$

1	$Z_2 \times A_5$	1	3	Normalizer of 5.
2	D_{12}	2	1	Normalizer of T.

J_3

$p = 2;\ |T| = 2^7;\ Q \cong QD$

1	A_5	4/1	1	$H;\Omega_4^-(2)$-module.
2	$Z_3 \times S_3$	4/2	1	Normalizer of $J(T)$. R Sylow in $L_3(4)$.

$p = 3;\ |T| = 3^5$

1	Z_8	2/1/2	1	$N_G(Z(T)) = N_G(T)$; T class 3.

J_4

$p = 2;\ |T| = 2^{21};\ Q \cong D^6$ 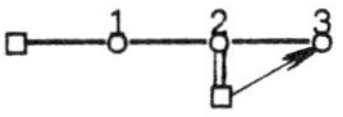

1	$Aut(M_{22})/Z_3$	12/1	1	M_{22}/Z_3-module.
2	$S_5 \times L_3(4)$	12/3	1	R special; $(L_2(4) \otimes L_3(2)^{(2)})/L_3(2)^{(1)}$.
3	M_{24}	11	1	Todd module.

$p = 3;\ |T| = 3^3$

1	$Aut(M_{22})/Z_3/D^6$	0	1	centralizer of involution.
2	$M_{24}/E_{2^{11}}$	0	2	2-parabolic
?	Existence and uniquness of maximal $U_3(3)$ and $PGL_3(4)$ left open.			

HS

$p = 2;\ |T| = 2^9;\ Q \cong Z_4 * D^2$

1	S_5	4/1/1	1	$Q/\Phi(Q)$ indecomposable; $Q/Z(Q)$ is $L_2(4)$-module.
2	$L_3(2)$	3/3	1	$R \cong (Z_4)^3$; split extension.

$p = 3;\ |T| = 3^2;$ 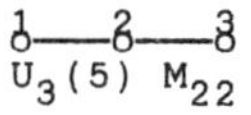

1	M_{22}	0	4	$G_1 \cap G_2 \cong (S_3 wr Z_2)/E_{16}$.
2	S_8	0	2	centralizer of outer automorphism.
3	$Z_2/U_3(5)$	0	2	$G_1 \cap G_2 \cong A_7$; $G_2 \cap G_3 \cong S_7$.
4	$Z_2/U_3(5)$	0	2	G_3 and G_4 fused in Aut(G).
5	$Z_2 \times Aut(A_6)$	0	1	centralizer of involution.
6	S_6/E_{16}	0	4	$G_1 \cap G_6 \cong A_6/E_{16}$; $G_2 \cap G_6 \cong D_8/E_9/E_{16}$
?	Existence and uniqueness of $F^*(M) \cong A_6, M_{11}, L_3(4)$ left in doubt.			

$p = 5;\ |T| = 5^3$

1	$Z_2/U_3(5)$	0	1	G_1 and G_2 are fused in Aut(G)
2	$Z_2/U_3(5)$	0	1	$G_2 \cap G_3 = N(T)$.

Mc

$p = 2;\ |T| = 2^7;$ 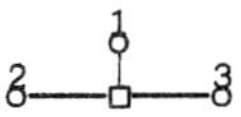

1	$L_4(2)$	1	1	centralizer of involution.
2	A_7	4	1	G_2 and G_3 are the normalizers of the two
3	A_7	4	1	members of $\mathfrak{A}(T)$ and are fused in Aut(G).
4	M_{22}	0	1	G_4 and G_5 are fused in Aut(G).
5	M_{22}	0	1	
6	$U_4(3)$	0	1	
7	$Z_2/L_3(4)$	0	1	

$p = 3;\ |T| = 3^6;\ Q \cong 3^{1+4}$ 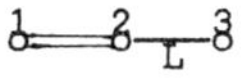

1	$Z_2/SL_2(5)$	4/1	1	H
2	M_{10}	4	1	$N_G(J(T))$; $\Omega_4^-(3)$-module.
3	$U_4(3)$	0	2	

$p = 5;\ |T| = 5^3$

1	Z_8/Z_3	2/1	1	H
2	$U_3(5)$	0	3	

Sz

$p = 2$; $|T| = 2^{13}$; $Q \cong Q^3$ 

1	$\Omega_6^-(2)$	6/1	1	$\Omega_6^-(2)$-module.
2	$S_3 \times L_2(4)$	8/2	1	R special; $(S_3 \otimes \Omega_4^-(2))/S_3$.
3	A_6/Z_3	6/4	1	R special; $A_6/Z_3//A_6$.

$p = 3$; $|T| = 3^7$ 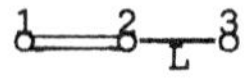

1	M_{11}	5	1	$N_G(J(T))$.
2	$Z_2/(SL_2(3)*D_8)$	4/2	1	$N_G(Z(T))$; R special; $(SL_2(3) \otimes D_8)/D_8$.
3	$Z_2/U_4(3)$	1	2	$(G_1 \cap G_3)/J(T) \cong M_{10}$.

$p = 5$; $|T| = 5^2$

1	$G_2(4)$	0	1	Contains $Aut(U_3(4))$; $G_1 \cap G_2 \cong A_5 \times A_5$.
2	$Z_2/(A_5 \times A_6)$	0	3	Contains normalizers of two 5's; Borel $\cong D_{10} \times A_5$
3	$Aut(HJ)$	0	1	$G_1 \cap G_3 = E(G_3)$; $G_2 \cap G_3 \cong Z_2/(D_{10} \times A_5)$
?	Existence and uniqueness of $F^*(M) \cong L_2(25)$ left open.			

Ly

$p = 2;\ |T| = 2^8$

1	A_{11}	1	1	Centralizer of involution.
2	$Aut(Mc)/Z_3$	0	1	Normalizer of 3.

$p = 3;\ |T| = 3^7$

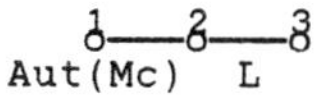

1	$Z_2 \times M_{11}$	5	1	$N_G(J(T))$.
2	$Z_2/(Z_8{*}SL_2(5))$	4/2	1	$N_G(Z(T))$.
3	Aut(Mc)	1	2	

$p = 5;\ |T| = 5^6;\ Q \cong 5^{1+4}$

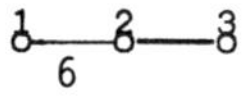

1	$L_3(5)$	3	1	$L_3(5)$-module.
2	$Z_2/(Z_4{*}SL_2(9))$	4/1	1	H
3	$G_2(5)$	0	1	

He

$p = 2;\ |T| = 2^{10};\ Q \cong D^3$

1	$L_3(2)$	(3+3)/1	1	H; $L_3(2)^{(1)} \oplus L_3(2)^{(2)}$.
2	S_6/Z_3	6	1	A_6/Z_3-module.
3	S_6/Z_3	6	1	G_2 and G_3 fused in Aut(G).

$p = 3;\ |T| = 3^3$

1	$S_6/Z_3/E_{64}$	0	2	G_1 and G_2 are fused in Aut(G).
2	$S_6/Z_3/E_{64}$	0	2	
3	$S_3/L_3(4)/E_4$	0	2	Normalizer of E_4.
4	S_7	1	1	Normalizer of Z(T).

$p = 5;\ |T| = 5^2$

1	$Z_2/Sp_4(4)$	0	3	Contains normalizer of two 5's.
2	$Z_4{*}SL_2(3)$	2	1	Normalizer of T.

$p = 7;\ |T| = 7^3$

1	$Z_3 \times S_3$	2/1	1	Normalizer of T.
2	$SL_2(7)$	2	3	

Ru

$p = 2; |T| = 2^{14}$

1	S_5	4/(1+1)/4/1	1	See lemma 12.12.
2	$L_3(2)$	8/3	1	R special.

$p = 3; |T| = 3^3$

1	$Aut(A_6)$	1	1	Normalizer of Z(T).
2	$^2F_4(2)$	0	1	Stabilizer of point in rank 3 representation.
3	$G_2(2)/E_{64}$	0	1	
?	Existence and uniqueness of maximal $U_3(3)$ left open.			

$p = 5; |T| = 5^3$

1	Z_4wrZ_2	2/1	1	Normalizer of T.
2	$GL_2(5)$	2	2	
3	$Z_2/U_3(5)$	0	2	

ON

$p = 2;\ |T| = 2^9$

1	$Z_2/L_3(4)$	1/1	1	Centralizer of involution.
2	$L_3(2)$	3/3	1	$R \cong (Z_4)^3$.

$p = 3;\ |T| = 3^4$

1	$Z_2/Z_5/QD$	4	1	Normalizer of T.
2	$Z_2/(Z_4 \times A_6)$	2	10	

$p = 7;\ |T| = 7^3$;

1	$Z_2/L_3(7)$	0	1	$G_1 \cap G_2 = N_G(T)$.
2	$Z_2/L_3(7)$	0	1	G_1 and G_2 fused in Aut(G).

Co_3

$p = 2$; $|T| = 2^{10}$;

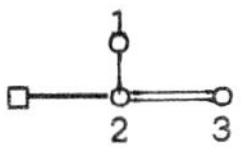

1	$Sp_6(2)$	1	1	Centralizer of involution.
2	$E_4/3^{1+2}$	6/2	1	$N_G(\Omega_1(Z_2(T)))$; R special.
3	$L_4(2)$	4	1	

$p = 3$; $|T| = 3^7$; $Q \cong 3^{1+4}$

1	$Z_2/(Z_4 * SL_2(9))$	4/1	1	H
2	$Z_2 \times M_{11}$	5	1	Normalizer of J(T).

$p = 5$; $|T| = 5^3$;

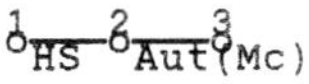

1	Aut(Mc)	0	1	$G_{1,3} \cong Z_2/U_3(5)$.
2	Aut($U_3(5)$)	0	1	$G_{1,2} = N_G(T)$.
3	HS	0	3	$G_{2,3} \cong Z_2/U_3(5)$; $G_{2,3} \notin (G_{1,3})^{G_3}$.

Co_2

$p = 2$; $|T| = 2^{18}$; $Q \cong D^4$; 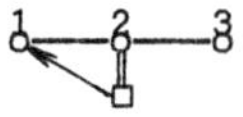

1	$Sp_6(2)$	8/1	1	H; spin module.
2	$S_3 \times S_5$	2/8/4	1	R special; Z(R) is $\Omega_4^-(2)$-module. $R/Z(R) = U \otimes S_3$ with U rank 5 indecomposable.
3	$Aut(M_{22})$	10	1	$N_G(J(T))$; dual Todd module.

$p = 3$; $|T| = 3^6$; $Q \cong 3^{1+4}$; 

1	Mc	0	4	$G_{1,3} = N_{G_1}(J(T))$; $G_{1,4} \cong U_4(3)$.
2	S_5/QD	4/1	1	H
3	$Aut(U_4(3))$	0	1	$G_{3,4} \cong Z_2/U_4(3)$.
4	$Z_2/U_6(2)$	0	2	$G_{1,4} \notin E(G_{3,4})^{G_4}$.

$p = 5$; $|T| = 5^3$; 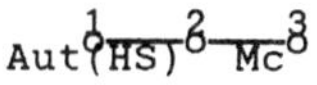

1	Mc	0	4	$G_{1,3} \cong U_3(5)$.
2	$Z_2/(Z_4{*}SL_2(3))$	2/1	1	H
3	Aut(HS)	0	3	

Co_1

$p = 2$; $|T| = 2^{21}$; $Q \cong D^4$; 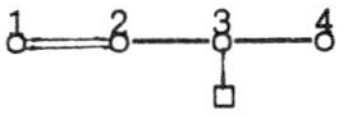

1	$\Omega_8^+(2)$	8/1	1	H; $\Omega_8^+(2)$-module.
2	$S_3 \times L_4(2)$	12/2	1	R special; $(S_3 \otimes \Omega_6^+(2))/S_3$.
3	$(S_6/Z_3) \times S_3$	12/4	1	R special; $((A_6/Z_3) \otimes S_3)/A_6$.
4	M_{24}	11	1	$N_G(J(T))$; dual Todd module.

$p = 3$; $|T| = 3^9$; $Q \cong 3^{1+4}$; 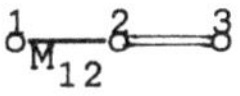

1	$GSp_4(3)$	4/1	1	H.
2	$GL_2(3)*GL_2(3)$	4/3	1	R special; $\Omega_4^+(3)/\Omega_3(3)$.
3	M_{12}/Z_2	6	1	$N_G(J(T))$.

$p = 5$; $|T| = 5^4$

1	$GL_2(5)$	2/1	1	H.
2	$Z_4 \times S_5$	3	1	Normalizer of J(T).

$p = 7$; $|T| = 7^2$

1	$Z_2/(L_3(2) \times A_7)$	0	4	Contains normalizer of two 7's.
2	$Z_3 \times SL_2(3)$	2	1	Normalizer of T.

M(22)

(2-locals of Aut(M(22))) $|T| = 2^{18}$; $Q \cong D^5$;

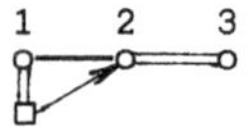

1	$O_6^-(2)$	1/8/1/1	1	H; $U_4(2)$-module.
2	$S_3 \times S_6$	8/1/4	1	R special; $(A_6 \otimes S_3)/1/\text{dual } A_6$.
3	$Aut(M_{22})$	10	1	Todd module.

$p = 3$; $|T| = 3^9$; $Q \cong 3^{1+6}$;

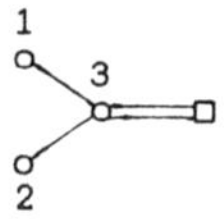

1	$P\Omega_7(3)$	0	1	G_1 and G_2 fused in Aut(G).
2	$P\Omega_7(3)$	0	1	$G_1 \cap G_2 = N_G(J(T)) \cong Z_2/\Omega_5(3)/E_35$.
3	$Z_2/E_9/E_{16}/E_8$	6/1	1	H.

$p = 5$; $|T| = 5^2$

1	$Aut(\Omega_8^+(2))$	0	1	Strongly 5-embedded in G.
2	${}^2F_4(2)'$	0	2	
3	S_{10}	0	3	G_3 and G_4 are fused in Aut(G).
4	S_{10}	0	3	

M(23)

$|T| = 2^{18}$

1	M(22)	1	1	Centralizer of a.
2	$Z_2/U_6(2)$	1/1	1	Centralizer of ab.
3	$Z_2/GU_4(2)$	8/2/1	1	Centralizer of abd.
4	M_{23}	11	1	$N_G(\langle D\cap T\rangle)$; Todd module.
5	$S_3 \times A_7$	8/6	1	R special; $(S_3 \otimes 4)/A_7$.

$p = 3$; $|T| = 3^{13}$; $Q \cong 3^{1+8}$; 1 2 3 o—o—o $S_3/P\Omega_8^+(3)$

1	$S_3/P\Omega_8^+(3)$	0	1	
2	$GL_2(3)/3^{1+2}/Q^3$	8/1	1	H.
3	$GL_3(3)$	3/3/1/3	1	R of class 4; $Z_3(R) \cong Z_3 \times$ Special; $GL_3(3)^{(1)}/L_3(3)^{(2)}/Z_2/L_3(3)^{(1)}$.

M(23)

$p = 5$; $|T| = 5^2$

1	S_{12}	0	3	D-subgroup.
2	$Sp_8(2)$	0	6	D-subgroup.
3	$S_3/P\Omega_8^+(3)$	0	2	D-subgroup.
4	$Aut(Sp_4(4))$	0	6	
5	$M(22)/Z_2$	0	1	Centralizer of a.
?	Existence of maximal M with $F^*(M) \cong L_2(25)$ left open.			

M(24)′

$p = 2$; $|T| = 2^{21}$; $Q \cong D^6$; 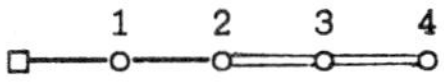

1	$Z_2/U_4(3)/Z_3$	12/1	1	H; $U_4(3)/Z_3$-module.
2	$L_3(2) \times A_6$	12/3	1	R special; $(L_3(2)^{(2)} \otimes A_6)/L_3(2)^{(1)}$.
3	$L_4(2) \times S_3$	8/6	1	R special; $(S_3 \otimes L_4(2))/\Omega_6^+(2)$.
4	M_{24}	11	1	Todd module.

$p = 3$; $|T| = 3^{16}$; $Q \cong 3^{1+10}$; 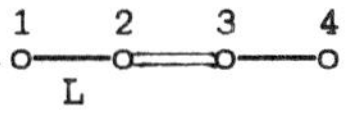

1	$\Omega_7(3)$	7	2	Natural module.
2	$GL_3(3)$	3/3/3/1/3	1	R of class 4; $Z_3(R) \cong Z_3 \times$ Special. $GL_3(3)^{(1)}/GL_3(3)^{(2)}/L_3(3)^{(2)}/Z_2/L_3(3)^{(1)}$.
3	$Z_2/(A_5 \times SL_2(3))$	8/4/2	1	R class 3; $Z_2(R) \cong E_3 6$. $(SL_2(3) \otimes A_5)/S_5/GL_2(3)$.
4	$Z_2/U_5(2)$	10/1	1	H.

M(24)′

(5-parabolics in M(24)) $p = 5$; $|T| = 5^2$

1	$S_5 \times S_9$	0	6	Normalizer of S_5; D-subgroup.
2	$Z_2 \times M(23)$	0	12	Centralizer of d.
3	$S_3 \times (S_3/P\Omega_8^+(3))$	0	4	Normalizer of S_3; D-subgroup.
4	$S_4 \times (S_3/\Omega_8^+(2))$	0	1	Normalizer of S_4; D-subgroup.
5	$O_{10}^-(2)$	0	12	D-subgroup.
6	$Aut(M(22))/E_4$	0	3	Centralizer of ab.
7	Aut(He)	0	24	Two classes in M(24)′ fused in M(24).

$p = 7$; $|T| = 7^3$

1	Aut(He)	0	1	G_1 and G_2 are fused in
2	Aut(He)	0	1	M(24); $G_1 \cap G_2 = N(T)$.

F_3

$p = 2$; $|T| = 2^{15}$; $Q \cong D^4$

1	A_9	8/1	1	H.
2	$L_5(2)$	5	1	Dempwolff group.

$p = 3$; $|T| = 3^{10}$

1	$GL_2(3)$	2/2/1/1/2/1	1	R class 4; R/R' indecomposable; $Z_3(R) = R' \cong E_9 \times S$, $|S| = 27$; S extraspecial exponent 9.
2	$GL_2(3)$	2/2/3/2	1	R class 3; R/R' indecomposable; $Z_2(R) = R' \cong E_{3^5}$.

$p = 5$; $|T| = 5^3$

1	$Z_2/(Z_4 * SL_2(3))$	2/1	1	Normalizer of T.
2	$GL_2(5)$	2	6	

$p = 7$; $|T| = 7^2$

1	$Aut(^3D_4(2))$	0	2	
2	$Z_3 \times (Z_2/SL_2(3))$	2	1	Normalizer of T.

F_5

$p = 2$; $|T| = 2^{14}$; $Q \cong D^4$

1	$O_4^+(4)$	8/1	1	H; $\Omega_4^+(4)$-module.
2	$L_3(2) \times Z_3$	6/2/3	1	R special; $(L_3(2)^{(1)} \otimes Z_3)/Z_3/L_3(2)^{(1)}$.

$p = 3$; $|T| = 3^6$; $Q \cong 3^{1+4}$

1	$Z_4 * SL_2(5)$	4/1	1	H.
2	$E_4/\Omega_4^+(3)$	4	1	$N_G(J(T))$.

$p = 5$; $|T| = 5^6$; $Q \cong 5^{1+4}$

1	$Z_4/Z_5/QD$	4/1	1	H.
2	$Z_4 * SL_2(5)$	2/1/2	1	R class 3; $R/Z(R) \cong 5^{1+2}$; $Z_2(R) \cong E_5 3$.

F_2

$p = 2;\ |T| = 2^{41};\ Q \cong D^{11};$

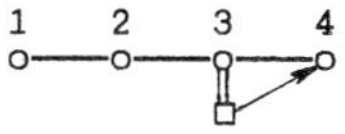

1	Co_2	22/1	1	H.
2	$S_3 \times Aut(M_{22})$	20/10/2	1	R class 3; $Z_2(R) \cong E_2 12$; $(S_3 \otimes \text{dual Todd})/\text{Todd}/S_3$.
3	$S_5 \times L_3(2)$	3/12/12/4/1/3	1	R has the following P-series: (with W an indecomposable $L_2(4)/1$): $((\text{dual}W)\otimes L_3(2)^{(1)})/(\Omega_4^-(2)\otimes L_3(2)^{(2)})/W/L_3(2)^{(1)}$.
4	$Sp_8(2)$	16/1/8	1	R special; Z(R) indecomposable $1/Sp_8(2)$.

$p = 3;\ |T| = 3^{13};\ Q \cong 3^{1+8};$

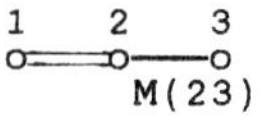

1	M(23)	0	1	$G_1 \cap G_2 \cong S_3/P\Omega_8^+(3)$.
2	$Aut(P\Omega_8^+(3))$	0	1	
3	$O_6^-(2)/Q^3$	8/1	1	H.

F_2

$p = 5$; $|T| = 5^6$; $Q \cong 5^{1+4}$; 1 o—2 o—3 o, Aut(F_5)

1	$Aut(F_5)$	0	1	
2	$S_5/(Z_4*D^2)$	4/1	1	H.
3	$L_3(5)$	3	1	

$p = 7$; $|T| = 7^2$; 1 o—2 o—3 o, ${}^2E_6(2)$ F_3

1	F_3	0	4	Centralizer of involution.
2	$Z_2/(F_4(2) \times E_4)$	0	2	Centralizer of involution.
3	$Z_2/{}^2E_6(2)/Z_2$	0	1	Centralizer of involution.
?	Existence and uniqueness of maximal M with $F^*(M) \cong L_2(49)$ left in doubt.			

F_1

$p = 2$; $|T| = 2^{46}$; $Q \cong D^{12}$; 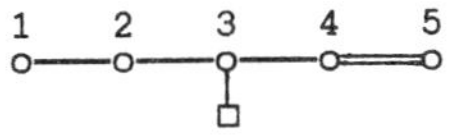

1	Co_1	24/1	1	H.
2	$S_3 \times M_{24}$	22/11/2	1	R class 3; $Z_2(R) \cong E_2 13$; $(S_3 \otimes \text{dualTodd})/\text{Todd}/S_3$.
3	$L_3(2) \times (S_6/Z_3)$	18/12/6/3	1	$((A_6/Z_3) \otimes L_3(2)^{(1)})/(A_6 \otimes L_3(2)^{(2)})/(A_6/Z_3)/L_3(2)^{(1)}$.
4	$L_5(2) \times S_3$	20/10/5	1	R class 3; $(L_5(2)^{(2)} \otimes S_3)/(L_5(2)^{(4)} \otimes S_3)/L_5(2)^{(1)}$.
5	$\Omega_{10}^{+}(2)$	16/10	1	R special; Z(R) is $\Omega_{10}^{+}(2)$-module.

$p = 3$; $|T| = 3^{20}$; $Q \cong 3^{1+12}$; 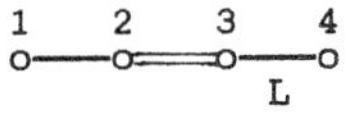

1	$Aut(Sz)/Z_2$	12/1	1	H.
2	$GL_2(3) \times M_{11}$	10/5/2	1	R class 3; $Z_2(R) \cong E_3 7$; $(GL_2(3) \otimes M_{11})/(Z_2 \times M_{11})/GL_2(3)$.
3	$L_3(3) \times SD_{16}$	6/6/2/3	1	R class 4; $Z_3(R) \cong E_9 \times$ special; $(SD_{16} \otimes L_3(3)^{(1)})/(D_8 \otimes L_3(3)^{(2)})/SD_{16}/GL_3(3)^{(1)}$.
4	$Z_2/\Omega_8^{-}(3)$	8	2	Natural module.

F_1

$p = 5$; $|T| = 5^9$; $Q \cong 5^{1+6}$; 1 2 3 o—o—o Aut(HJ)

1	$Z_2/((J_2/Z_2)*Z_4)$	6/1	1	H.
2	$S_3 \times GL_2(5)$	4/2/2	1	R class 3; $Z_2(R) \cong E_54$; $(S_3 \otimes GL_2(5))/(Z_4 \otimes S_3)/GL_2(5)$.
3	$Z_2 \times L_3(5)$	3/3	3	R is special.

$p = 7$; $|T| = 7^6$; $Q \cong 7^{1+4}$

1	$Z_2/((A_7/Z_2) \times Z_3)$	4/1	1	H.
2	$Z_2/(Z_3 \times SL_2(7))$	2/1/2	1	R class 3; $R/Z(R) \cong 7^{1+2}$; $Z_2(R) \cong E_73$.

$p = 11$; $|T| = 11^2$

1	$Z_2/(M_{12} \times L_2(11))$	0	12	
2	$Z_4/(L_2(11) \times L_2(11))$	0	6	
3	$Z_5 \times SL_2(5)$	2	1	Normalizer of T.

$p = 13$; $|T| = 13^3$

1	$Z_4/(SL_2(3) \times Z_3)$	2/1	1	H.
2	$Z_2/(Z_4*SL_2(13))$	2	6	

$U_4(3)$

$p = 2;\ |T| = 2^7;\ Q \cong D^2$ 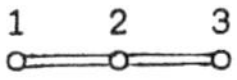

1	A_6	4	1	A_6-module
2	$S_3 \times S_3$	4/1	1	H.
3	A_6	4	1	A_6-module.

$L_4(3)$

$p = 2;\ |T| = 2^7;\ Q \cong D^2$

1	$PO_5(3)$	0	1	G_1 and G_2 fused in Aut(G).
2	$PO_5(3)$	0	1	

$G_2(3)$

$p = 2;\ |T| = 2^6;\ Q \cong D^2$

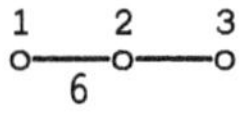

1	$L_3(2)$	3	1	
2	S_3	2/2/1	1	H.
3	$G_2(2)$	0	1	G_3 and G_4 are fused in Aut(G).
4	$G_2(2)$	0	1	

$P\Omega_8^+(3)$

$p = 2;\ |T| = 2^{12};\ Q \cong D^4$

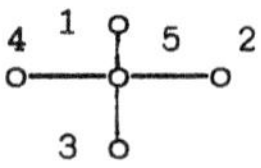

1,2,3,4	$\Omega_8^+(2)$	0	1	G_1, G_2, G_3, and G_4 are fused in Aut(G).
5	E_8/E_{81}	8/1	1	H

$U_6(2)$

$p = 3$; $|T| = 3^6$; $Q \cong 3^{1+4}$

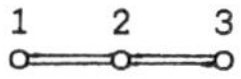

1	$Z_2/U_4(3)$	0	1	$G_1 \cap G_3 = N_G(J(T))$.
2	$S_3/(Q_8 \times Q_8)$	4/1	1	H.
3	$Z_2/U_4(3)$	0	1	G_1, G_3, and G_4 are fused in Aut(G).
4	$Z_2/U_4(3)$	0	1	

$F_4(2)$

$p = 3$; $|T| = 3^6$; $Q \cong 3^{1+4}$

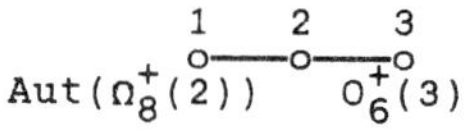

1	$O_6^+(3)$	0	1	$G_{1,3} = N_G(J(T))$.
2	$S_3/(Q_8 \times Q_8)$	4/1	1	H.
3	$Aut(\Omega_8^+(2))$	0	1	G_3 and G_4 are fused in Aut(G).
4	$Aut(\Omega_8^+(2))$	0	1	$G_3 \cap G_4 = N_G(J(T))$.

$^2E_6(2)$

$p = 3$; $|T| = 3^9$; $Q \cong 3^{1+6}$; 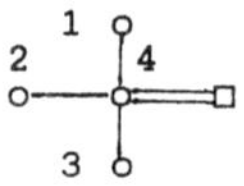

1	M(22)	0	1	Aut(G) is transitive on $\{G_1,G_2,G_3\}$.
2	M(22)	0	1	
3	M(22)	0	1	$G_1 \cap G_2 \cong P\Omega_7(3)$.
4	$Z_2/E_9/(Q_8 \times Q_8 \times Q_8)$	6/1	1	H.

$U_5(2)$

$p = 3$; $|T| = 3^5$. 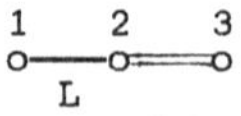

1	$\Omega_6^-(2)$	1	2	Stabilizer of nonsingular point in the natural module.
2	$Z_2 \times SL_2(3)$	2/1/1	1	Stabilizer of line.
3	S_5	4	1	Stabilizer of sum of 5 points; Normalizer of J(T).

2. Preliminary lemmas.

(2.1) Let p be a prime, G a group, and $Y = L_1 \times \ldots \times L_{p^e} \leq G$ with $L_i \cong L$ a nonabelian simple group. Assume either

(1) $p = 3$ and $G \cong F_3$, Sz, or Ly, or

(2) $p = 2$ and $G \cong$ Mc, Ly, Ru, ON, Co_3, M(22), or M(23).

Then $p^e = 2$, $G \cong$ M(22) or M(23), and $|Aut(Y)|_2 \leq 2^7$ or 2^{11}, respectively.

Proof. Let $r \in \pi(L)$. Then $m_r(G) \geq p^e$ and $|\pi(L)| \geq 3$. Because of the last observation we may choose $r > 3$ and because of the former the possibilities for r are limited. Let a be an element of order r in L_1. The structure of $C_G(a)$ can be found in Part I, Chapter 1, Section 5 of [23]. In particular reconciling this structure with the fact that a centralizes $L_2 \ldots L_{p^e}$ we conclude that $p^e = 2$, $r = 5$, and one of the following holds:

(a) $L \cong A_5$ and $G \cong$ M(22), Co_3, or Ru.

(b) $L \leq A_6$ and $G \cong$ M(23).

Actually if we look a little closer the cases $G \cong Co_3$ or Ru disappear. Namely in those cases a Sylow 5-group R of G is 5^{1+2} and without loss $S = R \cap Y$ is Sylow in Y. As $C_Y(a)$ is nonsolvable, a is not 5-central from the local strucure listed in [23]. Thus Z(R) is weakly closed in S and is not contained in L_i. These remarks are of course incompatible. This leaves $G \cong$ M(22) or M(23) and $L \cong A_5$ or A_6, which yields the bound on $|Aut(Y)|_2$.

(2.2) Let G be a group, Q a p-subgroup of G, and $E_{p^2} \cong E \leq Q$ such that for each e in $E^\#$, $\langle \mathcal{U}_{C_G(e)}(C_Q(e), p') \rangle = B$ is a p'-group independent of e. Then $\langle \mathcal{U}_G(Q, p') \rangle = B$.

Proof. This is because for each X in $\mathcal{U}_G(Q, 2')$,

$$X = \langle C_X(e) : e \in E^\# \rangle.$$

(2.3) Let G be a group, p a prime, $T \in Syl_p(G)$, $Q \le T$ such that $C_T(Q) \le Q$, and $z \in Z(T) \cap Q$ such that $Q \trianglelefteq C_G(z)$ and $z^{N(Q)} = z^G \cap Z(Q)$. Then Q is weakly closed in T with respect to G.

Proof. As $C_T(Q) \le Q \trianglelefteq T$, each p-element in $C_G(Q)$ is in Q. So if $Q \ne Q^g \le T$ then by hypothesis $z \in Z(Q^g)$. Then as N(Q) controls fusion of conjugates of z in Z(Q), Q^g is fused to Q in C(z). This contradicts the hypothesis that $Q \trianglelefteq C(z)$.

(2.4) Let V be a 5-dimensional vector space over F = GF(4) and G = SU(V) = $U_5(2)$. Let $T \in Syl_3(G)$. Then $T = X \times S$ where $S \cong Z_3 wr Z_3$, $C_V(S) = U$ is a nonsingular point of V, and $\mathcal{N}(T) = \mathcal{F} \cup \{G_1^r\}$ where $r \in N_G(T) - G_1$, $\mathcal{F} = (G_1, G_2, G_3)$ is a 3-basis for G, $X = Z(G_1) \cong Z_3$, and:

(1) $G_1 = N_G(U) \cong GU_4(2)$.

(2) $G_2 = N_G(W) \cong U_2(2) \times GU_3(2)$, where $W = C_V(Q)$ is a nondegenerate plane and $3^{1+2} \cong Q$.

(3) $G_3 = N_G(\theta) \cong S_5/E_{3^4}$, where θ is the set of points of V invariant under J(T), $N_G(\theta)$ induces S_5 on θ, and V is the orthogonal direct sum of the points in θ.

In particular $\Gamma(G,\mathcal{F})$ has the diagram listed in the introduction.

Proof. Let θ be a set of 5 nondegenerate points of V with V the orthogonal direct sum of the points of θ. Set $G_3 = N_G(\theta)$. Then the subgroup G_θ fixing θ pointwise is isomorphic to E_{81} and θ is the set of points invariant under G_θ. Further $G_3^\theta = S_5$. Thus G_3 contains a Sylow 3-subgroup of G which we may take to be T.

Let $U \in \theta$ and $G_1 = N_G(U)$. Then certainly $G_1 \cong GU_4(2)$. Notice $N_G(T) \le N_G(J(T)) = G_3$, so that $N_G(T)/T \cong E_4$ and T is contained in $|N_G(T):N_{G_1}(T)| = 2$ conjugates of G_1.

Let $S = T \cap G_1^\infty$ and $X = Z(G_1)$. Then $S \cong Z_3 wr Z_3$ and $T = X \times S$. In particular S contains a unique subgroup $Q \cong 3^{1+2}$. Notice $Q = C_T(W)$, where $W = U+U'$ for suitable $U' \in \theta$. In particular $G_2 = N_G(W) \cong U_2(2) \times GU_3(2)$.

Suppose $M \in \mathcal{N}(T) - \mathcal{E}$, where $\mathcal{E} = \mathcal{F} \cup \{G_1^r\}$. Now U,U′, and W are the irreducible FT-submodules of V and are nonisomorphic as FT-modules, so any proper nontrivial T-invariant subspace V_0 is the sum of these subspaces. In particular $N_G(V_0) \in \mathcal{E}$. Thus M is irreducible on V.

Similarly M is primitive on V. Namely if M is transitive on a set Σ of proper nontrivial subspaces with $V = \bigoplus_{P \in \Sigma} P$, then as dim(V) = 5 is prime, we conclude Σ is a set of 5 points. Thus T fixes two members of θ, which then must be U and U'. The remaining 3 points generate $W^{\perp}$ and as $\theta - \{U,U'\}$ is the only T-invariant triple of points in $W^{\perp}$, we have $\theta = \Sigma$. But then $M = G_3$, contrary to hypothesis.

If H is an irreducible subgroup of G and $K \trianglelefteq H$, then by Clifford's Theorem, V is a semisimple FK-module. Indeed as dim(V) = 5 is prime, if K is not irreducible then either K induces a group of scalars on V or V is the sum of 5 K-invariant points and K is a subgroup of the direct product of 5 copies of $F^{\#} \cong Z_3$. Recall also that $G = SU_5(2)$ contains no scalar transformations. Applying these remarks to H = M and recalling that M is primitive on V, we conclude each nontrivial normal subgroup of M is irreducible on V and that $O_3(M) = 1$. Then proceeding by induction on the order of a subnormal subgroup I of M and applying these remarks to $\langle I^M \rangle$, we conclude I is irreducible. This forces $F^*(M)$ to be a nonabelian simple group.

Let $D = X^M$. By 5.6 in [11], $\langle X,Y \rangle \cong E_9$ or $SL_2(3)$ for all $X \neq Y \in D$. Then as $F^*(M)$ is simple, the main theorem of [11] says $\langle D \rangle \cong PSp_n(3)$, $U_n(3)$, or $PGU_n(2)$. As $Z_3 \times (Z_3 wr Z_3) \cong T \in Syl_3(M)$ and $|M|$ divides the $|G|$, we have a contradiction.

(2.5) Let p be a prime, $G \cong SL_2(p)$, $X \in Syl_p(G)$, and V a GF(p)G-module upon which X acts quadratically and in which each chief factor is of dimension 2. Then V is semisimple and homogeneous.

Proof. Proceeding by induction on the number of chief factors, we may assume V is of dimension 4. Let U be an irreducible G-submodule. By hypothesis V/U is the natural GF(p)G-module so there is a point W in V with Z = [W,X] a point not in U. As X is quadratic on V, X centralizes Z. Thus W+Z is an X-invariant complement to U in V/P. Therefore V splits over U as a GF(p)G-module, completing the proof.

(2.6) Let G be J_4, z a 2-central involution in G, $H = C_G(z)$, $Q = O_2(H)$, $H^* = H/Q$, $X^* = O_3(H^*)$, and i an involution in H with $iz \in i^H$ and i inverting X^*. Then $C_H(i)$ is a $3'$-group.

Proof. Let $\tilde{H} = H/\langle z \rangle$. As i inverts X^*, i induces an outer automorphism on $E(H^*)/X^* \cong M_{22}$, so $C_{H^*}(i^*) \cong L_3(2)/E_{16}$ or $Z_4/Z_5/E_{32}$. We may assume 3 divides $|C_H(i)|$, so i^*

centralizes $L_3(2)/E_{16}$. Also as i inverts X^* and $\tilde{Q} = [\tilde{Q},X]$, $C_{\tilde{Q}}(i) = [\tilde{Q},i]$, so $C_{H^*}(i^*) = C_H(\bar{i})^*$. Further $C_H(\bar{i}) = O^2(C_H(\bar{i}))\langle i\rangle C_Q(i)$ and an element of order 7 in $C_H(i)$ fixes no hyperplane of $C_{\tilde{Q}}(i)$, so $C_{\tilde{Q}}(i) = C_Q(i)/\langle z\rangle$. But then $C_H(\bar{i}) = C_H(i)/\langle z\rangle$ so i is not fused to iz in H, contrary to hypothesis.

3. A generation lemma.

In this section G is a group, p is a prime, $M \leq G$ with $|G:M| \not\equiv 0 \bmod p$, Ω is a G-invariant set of p-subgroups of G, and $\Delta = M \cap \Omega$. For $D \in \Delta$ let $\mathfrak{D}(G,M,D)$ be the set of subgroups H of G with $D \leq H \not\leq M$, $|H:H\cap M| \not\equiv 0 \bmod p$, and $|\mathcal{N}_H(D)| = 1$. The following two results are established:

(3.1) If $G = \langle \mathcal{N}(D) : D \in \Delta \rangle$ then there exists $D \in \Delta$ such that

$$G = \langle N \in \mathcal{N}(D) : |N:N\cap M| \not\equiv 0 \bmod p \rangle$$

(3.2) $G = \langle \mathfrak{D}(G,M,D) : D \in \Delta \rangle$.

These results are proved when $p = 2$ in section 5 of [9]. Proofs for all p are included here for completeness.

First the proof of 3.1. Let $M \leq M_0 \in \mathcal{N}$. If the result holds for M_0 in place of M then there is $A \in M_0\cap\Omega$ and $N \in \mathcal{N}(A) - \{M_0\}$ with $|N:N\cap M_0| \not\equiv 0 \bmod p$. As M contains a Sylow p-group of G there is $g \in M_0$ with $|N^g:N^g\cap M| \not\equiv 0 \bmod p$. Choose $D \in N^g\cap M\cap\Omega$; then $G = \langle N^g, M_0 \rangle$, and the lemma holds.

So without loss $M = M_0 \in \mathcal{N}$. As $G = \langle \mathcal{N}(D) : D \in \Delta \rangle$ there is $D \in \Delta$ such that $\{M\} \neq \mathcal{N}(D)$. Let $N \in \mathcal{N}(D) - \{M\}$ be chosen so that $S \in Syl_p(M\cap N)$ is maximal. We may assume $D \leq S$ and $|N:N\cap M| \equiv 0 \bmod p$, so as M contains a Sylow p-group of G, $S < T \in Syl_p(M)$ and hence $S < N_T(S)$. So by maximality of S, $\mathcal{N}(N_T(S)) = \{M\}$. But as S is not Sylow in N, $N_N(S) \not\leq M$, so $S \trianglelefteq G$. M/S and N/S are distinct maximal subgroups of G/S containing $SD/S = 1$, so passing to G/S we may take $S = D = 1$, and $\Omega = \{1\}$.

Now by maximality of S, $M\cap K$ is a p'-group for each $K \in \mathcal{N} - \{M\}$. If K is itself a p'-group then $|K:M\cap K| \not\equiv 0 \bmod p$ and $G = \langle M,K \rangle$, so we are done. Thus p divides the order of K. In particular p divides the order of N and hence of G.

Let r be a prime divisor of $|G:M|$ and $R \in Syl_r(G)$. Then $R \neq G$ as p divides the order of G, so there is $I \in \mathcal{N}(R)$. We have shown p divides the order of I, so for some $g \in G$, p divides the order of $M\cap I^g$, and hence by the first sentence of the previous paragraph, $M = I^g$. But then M contains a Sylow r-group of G, whereas r

divides $|G:M|$, a contradiction. This completes the proof of 3.1.

Next the proof of 3.2. Assume the lemma is false and let $D \in \Delta$. Then G is not in $\mathfrak{D}(G,M,D)$, so by 3.1 there exists $A \in \Delta$ with

$$G = \langle N \in \mathcal{N}(A) : |N:N\cap M| \not\equiv 0 \bmod p\rangle.$$

By induction on the order of G,

$$N = \langle \mathfrak{D}(N,N\cap M,B) : B \in \Omega\cap N\cap M\rangle$$

for each $N \in \mathcal{N}(A)$ with $|N:N\cap M| \equiv 0 \bmod p$, so as $\mathfrak{D}(N,N\cap M,B) \subseteq \mathfrak{D}(G,M,B)$, the lemma holds.

4. A lemma on the representations of $SL_2(q)$.

The following result is established in section 4:

Lemma 4.1. Let $G = SL_2(q)$, $q = p^e$, p prime. Let V be a nontrivial irreducible $GF(p)G$-module and $T \in Syl_p(G)$. Assume $m(C_V(T)) = 1$ and $m([V,T]) \leq e+1$. Then one of the following holds:

(1) $e = 1$ and V is the natural module for G.

(2) $e = 1$ and V is the 3-dimensional $\Omega_3(p)$-module.

(3) $e = 2$ and V is the 4-dimensional $\Omega_4^-(p)$-module.

My original proof was somewhat longer. S. Smith suggested the use of 4.3 below, which simplifies things. In the remainder of this section assume the hypotheses of lemma 4.1. Let $F = GF(q)$ and $U = F \otimes_{GF(p)} V$. Then $U = \bigoplus_{b \in B} W^b$ for some irreducible FG-module W and a set B of coset representatives for $N_A(W)$ in $A = Aut(F)$. Observe $\dim_F(C_U(T)) = m(C_V(T)) = 1$ and $e+1 \geq m = \dim_F([U,T])$. But $C_U(T) = \bigoplus_{b \in B} C_{W^b}(T)$ and $C_{W^b}(T) \neq 0$, so as $\dim(C_U(T)) = 1$ we conclude:

(4.2) $U = W$ is irreducible and A acts on W.

(4.3) Let $g \in G - N_G(T)$. Then $W = [W,T] \oplus C_W(T^g)$, so $\dim(W) \leq e + 2$.

Proof. This is a consequence of [48].

Next $W = \bigotimes_{a \in A} W_{j(a)}^{\ a}$, where $(W_j : 1 \leq j \leq p)$ is a set of basic FG-modules and $1 \leq j(a) \leq p$. Recall we can choose notation so that $\dim(W_j) = j$ and W_2 is the natural FG-module. As A acts on W, $j(a) = j$ is constant for $a \in A$. Then $\dim(W) = j^e$.

By 4.3, $j^e \leq e+2$. It follows that either $j = 2$ and $e = 1$ or 2, or $j = 3$ and $e = 1$. In the last case 4.1.2 holds. If $e = 1$ and $j = 2$ then 4.1.1 holds. Finally if $e = 2$ and $j = 2$ then 4.1.3 holds, since $W_2 \otimes W_2^a$ is the $\Omega_4^-(p)$-module for $SL_2(p^2)$, where $\langle a \rangle = A$. Thus the proof of lemma 4.1 is complete.

5. Certain subgroups of symplectic groups.

In this section p is an odd prime, $q = p^e$ is a power of p, V is a symplectic space over $GF(q)$, and G is an irreducible subgroup of the extended symplectic group $GSp(V)$. Let Δ be the set of subgroups X of G of order p such that $\dim([X,V]) \leq 2$, and assume Δ is nonempty. The structure of G is determined under these hypotheses. Define

$$d = \min\{\dim([V,X]) : X \in \Delta\}$$

$$\Gamma = \{X \in \Delta : \dim([V,X]) = d\}$$

For $X \in \Delta$ the *root group* of X is

$$R(X) = \langle Y \in \Delta : [V,Y] = [V,X]\rangle.$$

Define $\Omega = \{R(X): X \in \Gamma\}$, let $(\Omega_i : 1 \leq i \leq n)$ be the orbits of $H = \langle\Gamma\rangle$ on Ω and let $H_i = \langle\Omega_i\rangle$ and $V_i = [H_i,V]$.

The following two results are established:

Theorem 5.1. Assume $q = p$. Then either H is the central product of the groups (H_i: $1 \leq i \leq n$), or the following hold:

(1) $p = 3$.

(2) $H = PM$ where P is an elementary abelian 3-group, $M = M_1 \times \ldots \times M_m$, $M_i \cong Q_8$, and $[M,A] = M_i \times M_j$ for each $A \in \Gamma$.

(3) V is the orthogonal direct sum of the subspaces $[V,M_i]$, each of which is a plane.

(4) G permutes the groups (M_i: $1 \leq i \leq m$) transitively.

Theorem 5.2. Assume $q = p$ and H is the central product of the groups (H_i: $1 \leq i \leq n$). Then

(1) G transitively permutes (H_i: $1 \leq i \leq n$).

(2) V is the orthogonal direct sum of the subspaces V_i, $1 \leq i \leq n$.

(3) $H_i \leq Sp(V_i)$ and $N_G(V_i)$ acts irreducibly on V_i.

(4) Either H_i acts irreducibly on V_i or V_i is the direct sum of subspaces V_{i1} and V_{i2} of equal dimension and either

(a) $H_i \cong SL(V_{i1})$ acts naturally on V_{i1} with V_{i2} the dual of V_{i1} for H_i. V_{ij} is totally singular.

(b) V_i is the orthogonal direct sum of V_{i1} and V_{i2} and either $H_i \cong Sp(V_{i1})$ acts naturally on V_{ij}, or $p = 3$, $\dim(V_i) = 4$, and $H_i \cong Z_3/(Q_8 \times Q_8)$.

Notice that when $q = p$, Theorems 5.1 and 5.2 reduce us to the case $G = \langle\Gamma\rangle$ transitive on Ω, which is handled in Theorem 5.3.

Theorem 5.3. Assume $G = \langle\Gamma\rangle$ and G is transitive on Ω. If $q \neq p$ assume $F^*(G)/Z(G)$ is simple and the representation of $F^*(G)$ on V is absolutely irreducible and defined over no proper subfield of $GF(q)$. Then one of the following holds:

(1) $d = 1$ and $G = Sp(V)$.

(2) $q = p$, $d = 2$, and $G \cong Sp_m(p^2)$ or $SU_m(p)$ acts naturally on V and is determined up to conjugacy in $Sp(V)$ by its isomorphism type, where $m = \dim(V)/2$.

(3) $q = 3$, $d = 2$, $\dim(V) = 4$, and $G \cong A_5/(Q_8 * D_8)$ or $SL_2(5)$ is determined up to conjugacy in $Sp(V)$.

Theorems 5.1 through 5.3 are stronger results than necessary for the applications in this paper, but the results seem nice and are probably useful elsewhere. Thus I have chosen to state and prove them in this generality, but will occassionally omit a few details in the proof.

The proof involves a series of lemmas. As G is irreducible on V, certainly:

(5.4) $O_p(G) = 1$.

Thus for $A \in \Gamma$ there exists $B \in \Gamma$ with $\langle A,B\rangle$ not a p-group. Denote by $\theta(A)$ the set of all such subgroups $\langle A,B\rangle$. Throughout this section let $A,B \in \Gamma$ and $K = \langle A,B\rangle \in \theta(A)$.

(5.5) Let $X \in \Delta$. Then

(1) $[X,V]$ is totally singular.

(2) X is quadratic on V; that is $[V,X,X] = 0$.

(3) $[X,V]^\perp = C_V(X)$.

Proof. If $[X,V]$ is not totally singular then as $\dim([X,V]) \leq 2$, $[X,V]$ is nondegenerate of dimension 2, so $V = [V,X]\oplus[V,X]^{\perp}$. But then $[V,X] = [V,X,X] < [V,X]$, a contradiction. So (1) holds.

Suppose $Z = [V,X,X] \neq 0$, and let U be a point in $V-Z^{\perp}$. $[V,X] \leq W = U+[V,X]$, so X acts on W. Thus X acts on $W_0 = W^{\perp}\cap[V,X]$. Notice $W_0 = U^{\perp}\cap[V,X]$ by (1). But W_0 is a point distinct from Z, while Z is the unique X-invariant point of $[V,X]$, a contradiction.

Part (3) is 22.1 in [6].

(5.6) (1) For each $D \in \Gamma$, either $[A,D] = 1$, $\langle A,D\rangle \in \theta(A)$, or $\langle A,D\rangle \cong p^{1+2}$.

(2) Either $K \cong SL_2(p^f)$ for some f dividing de or $p = 3$, $d = 2$, and $K \cong SL_2(5)$ or $Z_3\times SL_2(3)$.

(3) Let t be the involution in K. Then $[V,t]$ is a nondegenerate subspace of V and either $[V,K] = [V,t] = [V,A]\oplus[V,B]$ or $d = 2$, $p = 3$, $\dim([V,t]) = 2$, $K \cong SL_2(3) \times Z_3\epsilon$, and $[V,t] = [V,O^3(K)]$.

(4) One of the following holds:

(i) $K \cong SL_2(p^f)$ acts naturally on $[V,t]$.

(ii) $K \cong SL_2(p^f)$ with f dividing e, $\dim([V,t]) = 4$, and $[V,t]$ is the sum of two natural modules for K.

(iii) $p = 3$ and $K \cong SL_2(5)$ or $Z_3 \times SL_2(3)$.

(5) The centralizer in $Sp([V,t])$ of K contains no Q_8-subgroup.

Proof. Parts (1) and (2) are a consequence of 5.5.2 and Ho [47]. Moreover it is well known that if $d = 1$ then (3)-(5) hold, so we may take $d = 2$. Let $W = [V,t]$. As t is a p'-element, W is nondegenerate. Thus $\dim(W)$ is even and as $K = \langle A,B\rangle$ and $C_W(K) \leq C_W(t) = 0$, $W = [W,A]\oplus[W,B]$ is of dimension 2 or 4. In the latter case (3) holds. In the former A induces transvections on $W^{\perp}$, so $K/C_K(W^{\perp}) \cong SL_2(r)$ or Z_p, for some power r of p, by (2). Then (2) forces $K \cong SL_2(3) \times Z_3\epsilon$ and $W = [V,O^3(K)]$. So (3) is established. To prove (4) and (5) we may take K faithfully embedded in $Sp(V) \cong Sp_4(q)$ by (3), after which (4) and (5) follow from the structure of this 4-dimensional symplectic group.(cf. [30] and [2].)

(5.7) Let $\langle t\rangle = O_2(Z(K))$ and $J = \langle K^{C(t)}\rangle$. Then one of the following holds:

(1) $J \cong SL_2(p^f)$, with f dividing de, and $[V,t] = [V,J]$.

(2) $p = 3$, $J \cong SL_2(5)$, $Z_3/(Q_8\times Q_8)$, or $A_5/(Q_8{*}D_8)$ and $[V,t] = [V,J]$.

(3) $p = 3$, $\dim([V,t]) = 2$, $Q_8 \cong J_0 = \langle O^3(K)^{C(t)}\rangle$, and $[V,t] = [V,O^3(J_0)]$.

Proof. Let $U = [V,t]$. If U is of dimension 2 then it is easy to see that (1) or (3) holds from 5.6, so we can assume $d = 2$ and $\dim(U) = 4$ by 5.6. Then J is faithfully embedded in $Sp(V) \cong Sp_4(q)$, and then the structure of this latter group says (1) or (2) holds.(cf. [30]. J cannot be A_7/Z_2 when $p = 7$ or $Z_5/(Q_8{*}D_8)$ or $A_5/(Q_8{*}D_8)$ when $p = 5$ because of 5.6.1 and 5.6.2.)

Let $\alpha(A)$ consist of the groups $\langle X^{C(t)}\rangle$, where $X \in \theta(A)$, $A \in \Gamma$, $\langle t\rangle = O_2(Z(X))$, and $\dim([V,t])$ is minimal subject to these constraints. Let $\Sigma(A)$ consist of those J in $\alpha(A)$ with $[A,V] \leq [V,t]$ together with the groups $\langle O^3(X)^{C(t)}\rangle$ such that $[A,V] \not\leq [V,t]$. Let $\Sigma = \bigcup_{A\in\Gamma} \Sigma(A)$, $M = \langle\Sigma\rangle$, $(\Sigma_i : 1 \leq i \leq m)$ the orbits of M on Σ, $M_i = \langle\Sigma_i\rangle$, and $U_i = [V,M_i]$.

(5.8) Let $J \in \Sigma(A)$ and $\langle t\rangle = Z(J)$. Then

(1) $J \cong SL_2(p^f)$, with f dividing de, or $p = 3$ and J is $SL_2(5)$ or Q_8.

(2) If $t \in L \in \Sigma$ then $L = J$; indeed J centralizes no Q_8-subgroup containing t.

Proof. Part (1) is essentially immediate from 5.7. Notice the minimality condition on $\dim([V,t])$ excludes the second two possiblities for J in 5.7.2. If $t \in L \in \Sigma-\{J\}$ then as $J/\langle t\rangle$ and $L/\langle t\rangle$ have just one C(t)-chief factor, $[J,L] = 1$. This is impossible since from 5.6.5, J centralizes no Q_8-subgroup of $Sp([V,t])$.

Lemma 5.8 allows us to apply the results of [10]. (See also section 9.) In particular 5.8.2 says that (in the notation of [10],) A(J) is empty for $J \in \Sigma$, so by Theorem 4 of [10], M is the central product of the groups $(M_i : 1 \leq i \leq m)$, and hence M_i is transitive on Σ_i.

Let's see next that $[U_i,M_j] = 0$ for $i \neq j$. For if not there is $J \in \Sigma_i$ and $L \in \Sigma_j$ with $[V,J,L] \neq 0$. Then as $[V,L] = [V,Z(L)]$, L is faithful on $[V,J]$, forcing $Z(L) = Z(J)$ and contradicting 5.8.2.

So $[U_i,M_j] = 0$ for $i \neq j$. Hence as G is irreducible on V, $V = \bigoplus_{i=1}^{m} U_i$ with G transitively permuting $(M_i : 1 \leq i \leq m)$ and $N_G(U_i)$ irreducible on U_i. The last remark shows U_i is nondegenerate or totally singular. As $[V,J]$ is nondegenerate and contained in U_i, U_i is nondegenerate and V is the orthogonal direct sum of $(U_i : 1 \leq i \leq m)$.

Notice if $A \leq J \in \Sigma(A)$ then $R(A) \in Syl_p(J)$ and hence after suitable relabeling, $H_i = M_i$. By 5.8, $J \cong SL_2(r)$, where either $r = |R(A)|$ or $p = 3$ and $r = 5$. Then by Theorems 1 and 2 in [10], either H_i is a group $G(r)$ of Lie type over the field of order $r = |R(A)|$ and $R(A)$ is a long root subgroup of H_i, or $r = 3$ and $H_i \cong M_{11}$, or $p = 3$, $J \cong SL_2(5)$, and H_i is of Lie type over GF(5). The second case is out by 5.5.2 as there are conjugates of A in M_{11} generating an A_4-subgroup. In the third case, either $H_i = J$ or $A = Z(P)$ for some subgroup $P \cong 5^{1+2}$ of H_i. But if P exists then $m([V,A]) \geq 5$ as 5 is the minimal dimension of a faithful FP-module.

Notice that we have just shown that if $A \leq J$ then Theorem 5.1 holds, that the first three parts of Theorem 5.2 hold, and that $H_i = M_i$ is of Lie type over GF(r), where $r = |R(A)|$ and $R(A)$ is a long root group of H_i. On the other hand suppose $A \not\leq J$. Then by 5.6.3, $\dim([V,J]) = 2$, so J is irreducible on [V,J] and thus M_i is irreducible on U_i. Hence $Z(M_i)$ is cyclic. Further if $1 \neq tt^g = e \in Z(M_i)$ for some $g \in M_i$ then $U_i = [V,e]$ as M_i is irreducible on U_i. Thus $U_i = [V,t] \oplus [V,t^g]$ is of rank 4 so $M_i \leq Sp_4(q)$ and then from the structure of that group, $t \in O_2(M_i)$ as $d = 2$. It follows from Theorem 1 in [10] and 5.8 that either M_i is of Lie type over GF(3) or $t \in O_2(M_i)$, and if M_i is of Lie type then either $AJ \trianglelefteq C(t)$ and A is a long root group of M_i or $A = \langle xy \rangle$ with $SL_2(3) \cong \langle x \rangle J \trianglelefteq C(t)$ and $y \in C(\langle x \rangle J)$. In the former case replace J by AJ. In the latter case $\langle x \rangle \in \Gamma$ and we argue on $\langle x \rangle J$ in place of J; here we get a contradiction from $\dim([V,t]) = 2 = d$. Namely applying Theorems 5.2 and 5.3 to $\langle x \rangle$ and using the fact that [V,t] is of dimension 2, we conclude that [V,x] is of dimension 1. But this contradicts $d = 2$.

So assume $t \in O_2(M_i)$. Then either $t \in Z(M)$, so $M_i = J$, or $[O_2(M_i),A] \neq 1$. In the first case we have shown that parts (1), (3) and (4) of Theorem 5.1 hold. As $d = 2$, A fails to centralize at most two of the M_i. As $[A,V] \leq [V,t]$, A fails to centralize at least two of the M_i. Further H/M is an abelian 3-group as A, and hence also H, acts on $[V,M_i]$ for each i. Therefore Theorem 5.1 holds in this case.

In the second case there is a characteristic subgroup Q of $O_2(M_i)$ of class at most 2 with A faithful on Q. Then there is $x \in Q$ with $L = \langle A,A^x \rangle \cong SL_2(3)$. As Q is of class at most 2, $\langle s \rangle = \Phi(O_2(L)) \leq Z(Q) \leq Z(M_i)$, and as M_i is irreducible on U_i, $U_i = [V,s]$. By 5.6, V is of dimension 2 or 4 and as $d = 2$ it is the latter. Thus $M_i \leq Sp_4(3)$ with $O_2(M_i) \not\leq Z(M_i)$, so the structure of $Sp_4(3)$ shows that $M_i \cong A_5/(Q_8{*}D_8)$ satisfies 5.3.3 on U_i. Further Theorem 5.2 holds with $H_i = M_i$.

To summarize we have reduced to the case where $M_i = H_i$ is of Lie type and R(A) is a long root group of H_i and Theorems 5.1 and 5.2 hold modulo the verification of 5.2.4.

Let's now verify 5.2.4. As $N_G(V_1)$ is irreducible on V_1, $V_1 = \bigoplus_{j=1}^{m} V_{1j}$ with V_{1j} an irreducible H_1-module. $[A,V_{1j}] \neq 0$ for each j, so $m \leq 2$ and if $m = 2$ then $d = 2$ and A induces transvections on V_{1j}. Now appeal to a result of McLaughlin [28] to conclude that $H_1/C_{H_1}(V_{1j}) \cong Sp(V_{1j})$ or $SL(V_{1j})$. If $\dim(V_{1j}) > 2$ then V_{1j} does not admit $SL(V_{1j})$ if V_{1j} is nondegenerate, so the space is totally singular. Now 5.2.4.a holds by 5.4 in [2]. If $\dim(V_{1j}) = 2$ then $SL(V_{1j}) = Sp(V_{1j})$ so we may assume our group is symplectic. If V_{1j} is nondegenerate it is an easy exercise to show 5.2.4.b holds, so assume the space is totally singular. If H_1 is homogeneous on V_1 then 10.2 and 10.3 in [2] imply there exists a nondegenerate irreducible I for H_1, so replacing V_{11} and V_{12} by I and $I^{\perp}$, we are reduced to a previous case. On the otherhand by 5.4 in [2] the stabilizer in $Sp(V_1)$ of V_{11} and V_{12} is $L \cong GL(V_{11})$, so as L is transitive on its subgroups acting as $Sp(V_{11})$ on V_{11}, and as there exists such a subgroup homogeneous on V_1, all such subgroups are homogeneous, and the proof is complete.

So Theorem 5.2 is established. Hence in the remainder of this section we assume the hypotheses of Theorem 5.3. Moreover we have shown that $G = G(r)$ is a group of Lie type over the field of order $r = p^f$, where f divides de, and $R = R(A)$ is a long root group of G. In particular $|R| = r$ and R is in the center of a Sylow p-group T of G. Let $U = [V,A]$; notice also $U = [V,R]$ by 5.6.4. Set $F = GF(q)$.

(5.9) The representation of G on V can be writen over no proper subfield of F.

Proof. This holds by hypothesis if $q > p$ and is trivial if $q = p$.

(5.10)(1) $E = GF(r)$ or $GF(r^2)$ is a splitting field for G, for G an ordinary or twisted group, respectively.

(2) $F \leq E$.

Proof. The first remark is well known. (For example it is implicit in [21] that a splitting field for a Cartan group of G is a splitting field for G.) The second is clear if $q = p$ so assume $q > p$. Then by hypothesis V is an absolutely irreducible FG-module. Let k be the field generated by E and F and $V^k = k\otimes_F V$. As V is absolutely irreducible so is V^k. By 26.6 in [6] there is a smallest subfield L of k such that the representation of G on V^k can be written over L; by 5.9, L = F. Hence by (1), $F \leq E$.

(5.11) If V is an absolutely irreducible FG-module then $\dim(C_V(T)) = 1$.

Proof. This is because G is of Lie type. (cf. [21].)

(5.12) $C_\Gamma(A) \leq C_G(U)$.

Proof. Let $D \in C_\Gamma(A)$ and suppose $0 \neq [U,D]$. Then d = 2 and by 5.5, U+[V,D] = W is of dimension 3 and not totally singular, so W is contained in a nondegenerate 4-space Z. By 5.5.2, X = AD centralizes $Z^\perp$, so $X \leq Sp(Z)$. But then as d = 2, $O^{p'}(C_{Sp(Z)}(A)) \leq C(U)$.

(5.13) $U \leq C_V(T)$.

Proof. Let $M = \langle C_\Gamma(A)\rangle$. As G is of Lie type and R is a long root group of G, either $T \leq M$ or $G/Z(G) \cong G_2(r)$, ${}^3D_4(r)$, or $U_3(r)$. By 5.8.2 $G_2(r)$ and ${}^3D_4(r)$ are out, so by 5.12 we may take $G/Z(G) \cong U_3(r)$. If U is a point the lemma is clear so take U to be a line. Let t be the involution in K. Then T = [T,t] so if U = [U,t] the lemma holds. Hence we may assume $U \not\leq [V,t]$, so by 5.6.3 we may take $K \cong SL_2(3)$ and [V,t] a line. There is $x \in t^G$ with $X = \langle x,t\rangle \cong E_4$. Then $\langle x\rangle K \cong Z_4 * SL_2(3)$ is faithful on [V,t] so $[V,t,x] \neq 0$ and hence [V,X] is of dimension 3. As [V,t] and [V,x] are nondegenerate, each is a complement to the radical W of [V,X]. Thus [X,W] = 1. This contradicts $W \leq [V,X]$.

(5.14) If V is an absolutely irreducible FG-module the Theorem holds.

Proof. Assume V is absolutely irreducible. By 5.11 and 5.13, U is a point. Hence R is contained in the root group of a transvection so if $\langle R,R^g\rangle$ is not a p-group it is in $\Sigma(A)$. But the only groups of Lie type in which a long root group and any conjugate generate an abelian group or $SL_2(r)$ are coverings of $U_m(r)$ and $PSp_m(r)$. Hence $G/Z(G) \cong U_m(r)$ or $PSp_m(r)$. Also f divides e, so by 5.10, either r = q or G is unitary and $q = r^2$. In the first case G = Sp(V) by McLaughlin [28]. In the second there exists $S \in R^G$ with $\langle K,S\rangle = L \cong U_3(r)$ or $SU_3(r)$. As L is generated by three conjugates of R, $\dim([V,L]) \leq 3$, so as 3 is the minimum dimension of a nontrivial FG-module, [V,L] is irreducible of dimension 3. As $[V,K] \leq [V,L]$, [V,L] is not totally singular so [V,L] is nondegenerate, contradicting [V,L] of odd dimension.

If $q > p$ then by hypothesis V is absolutely irreducible. So by 5.14 we may take $q = p$ and V not absolutely irreducible. Let $k = \mathrm{End}_{FG}(V)$, so that $k \neq F$. $|k:F| \leq d \leq 2$, so $d = 2 = |k:F|$. Then r divides p^2 while $k \leq E$, so by 5.10, either $r = p^2$ or G is twisted and $r = p$. Now regarding V as a kG-module, R is contained in the root group of a k-transvection so $G/Z(G) \cong Sp_m(p^2)$ or $U_m(r)$ by an argument in the preceding paragraph. If $r = p^2$ then R is the root group of a k-transvection, so by McLaughlin [28], $n = 2m$ and $G \cong Sp_m(r)$ acts naturally on V. By [2], G is determined up to conjugacy in Sp(V). So take $r = p$; then $G/Z(G) \cong U_m(p)$ and as $\mathrm{End}_{FG}(V) = GF(p^2)$, by [2] G is contained in a subgroup G_0 of Sp(V) isomorphic to $SU_{n/2}(p)$ or $Sp_{n/2}(p^2)$ acting naturally on V, and determined up to conjugacy in Sp(V). The case $G_0 = Sp_{n/2}(p^2)$ is out by 5.14. Thus $G_0 = SU_{n/2}(p)$, so G is generated by long root groups of G_0 and is irreducible on V. Therefore $G = G_0$. (cf. 6.6.)

The proof of the three main theorems of this section is now complete.

6. Groups generated by transvections.

In this section V is an n-dimensional vector space over a finite field F of characteristic p and G is an irreducible subgroup of Δ = GL(V), Sp(V), or GU(V), such that the set R(G) of root groups of transvections in Δ is nonempty.

In section 6, G and its action on V are determined under these hypotheses and a few corollaries are derived. Recall that if G = <R(G)>, G is irreducible on V, and Δ is GL(V) or Sp(V), then G is known by McLaughlin [28,29]. This result of Mclaughlin is used heavily.

For $X \in R(G)$ let $\theta(X)$ consist of those $Y \in R(G)$ such that $\langle X,Y\rangle$ is not a p-group. Let R_i, $1 \leq i \leq r$, be the orbits of $\langle R(G)\rangle = H$ on R(G), $H_i = \langle R_i\rangle$, and $V_i = [V,H_i]$. Recall:

(6.1) Let $X \in R(G)$. Then

(1) $X \cong E_q$, where $q = |F|$ if Δ = GL(V) or Sp(V) and $q^2 = |F|$ if Δ = GU(V).

(2) $\dim([V,X]) = 1$, $[V,X] = [V,x]$, and $C_V(X) = C_V(x)$ for each $x \in X^{\#}$.

(3) If $Y \in \theta(X)$ then $K = \langle X,Y\rangle \cong SL_2(q)$, $\dim([V,K]) = 2$, K is irreducible on [V,K], $V = [V,K]\oplus C_V(K)$, and [V,K] is nondegenerate if Δ = Sp(V) or GU(V).

(4) If $Y \in R(G) - \{X\}$ and $Y \notin \theta(X)$ then either $\langle X,Y\rangle = X \times Y$ or Δ = GL(V), $\langle X,Y\rangle$ is special of order q^3, and $[X,Y] \in R(G)$.

As G is irreducible on V, $O_p(G) = 1$, so by the Baer-Suzuki Theorem (cf. 39.6 in [6]):

(6.2) $\theta(X)$ is nonempty for each $X \in R(G)$.

(6.3) If $X,Y \in R(G)$ and $Z = [X,Y] \in R(G)$ then, interchanging the roles of X and Y if necessary, $C_V(X) = C_V(Z)$ and $[Y,V] = [Z,V]$.

Proof. By 6.1.4, $E = \langle X^Y\rangle = X \times X^y \cong E_{q^2}$ for $y \in Y^{\#}$. If $[V,X] = [V,X^y]$ then $[V,X] = [V,E] = [V,Z] \leq C_V(Y)$, as Y acts on the point [V,Z]. So X acts on $C_V(Y)$. Hence $C_V(Y)$ =

$C_V(\langle Y^X\rangle) = C_V(Z)$. If $[V,X] \neq [V,X^Y]$ then $[V,E] = [V,X]\oplus[V,X^Y]$ and $[V,Y] = [V,E,Y] = C_{[V,E]}(Y) = [V,Z]$. Therefore $[V,Y] = [V,YZ] = [V,Y^x]$ for $x \in X^{\#}$, and we are reduced to an earlier case.

(6.4) If $i \neq j$ then $[H_i,H_j] = 1$.

Proof. We may assume $X \in R_i$ and $Y \in R_j$ with $[X,Y] \neq 1$. If $Y \in \theta(X)$ then Y is conjugate to X in $\langle X,Y\rangle \leq H \leq N(R_i)$ by 6.1, contradicting $R_i \neq R_j$. So by 6.1.4, $[X,Y] = Z \in R(G)$. Let $Z \in R_k$. We show $k = i$; then by symmetry $i = k = j$, a contradiction.

So assume $k \neq i$. By 6.3, either $[X,V] = [Z,V]$ or $C_V(X) = C_V(Z)$. We consider only the first case; the second is handled via a dual argument. By 6.2 there is $X_1 \in \theta(X)$. As we observed above, $X_1 \in R_i$. Let $K = \langle X,X_1\rangle$. By 6.1.3, X_1 does not centralize $[V,X] = [V,Z]$, so $[Z,X_1] \neq 1$. As $k \neq i$, $X_1 \notin \theta(Z)$. Thus by 6.3, and as $[V,Z] = [V,X] \neq [V,X_1]$, we have $C_V(Z) = C_V(X_1)$. But there is $X_2 \in R(K) - \{X,X_1\}$ and by symmetry between X_1 and X_2, $C_V(Z) = C_V(X_2)$. This is impossible since by 6.1.3, $C_V(X_1) \neq C_V(X_2)$.

(6.5) (1) H is the direct product of the groups H_i, $1 \leq i \leq r$, which are permuted transitively by G.

(2) V is the direct sum of the subspaces V_i, $1 \leq i \leq r$, which are permuted transitively by G. If $\Delta = Sp(V)$ or $GU(V)$ then the sum is orthogonal.

(3) H_i is irreducible on V_i and transitive on R_i.

Proof. By 6.4, H is the central product of the groups H_i, $1 \leq i \leq r$. Thus as H is transitive on R_i, so is H_i. As H_i centralizes H_j for $i \neq j$, X centralizes $[V,Y]$ for $X \in R_i$ and $Y \in R_j$. Hence $H_i = \langle R_i\rangle$ centralizes $\langle [V,Y] : Y \in R_j\rangle = V_j$. Let $U_i = \langle V_j : j \neq i\rangle$. As G is irreducible on V, $V = [V,H] = U_i+V_i$ and $C_V(H) = 0$. As $[H_i,V_j] = 0$, $U_i \cap V_i \leq C_V(H) = 0$. Therefore $V = \bigoplus_{i=1}^{r} V_i$. Then as G is irreducible on V, G is transitive on the spaces V_i and on the groups H_i, $1 \leq i \leq r$, and $N_G(V_i)$ is irreducible on V_i. Thus V_i is a semisimple H_i-module. So as $\dim([V,X]) = 1$ for $X \in R_i$, H_i is even irreducible on V_i.

Finally if $\Delta = Sp(V)$ or $GU(V)$ then as H_i is irreducible on V_i, either V_i is totally singular or nondegenerate. Since $[V,K] \leq V_i$ and $[V,K]$ is nondegenerate for $K = \langle X,Y\rangle$, $X \in R_i$, and $Y \in \theta(X)$, we conclude V_i is nondegenerate. Thus V is the orthogonal direct sum of the V_i, $1 \le i \le r$.

(6.6) Assume $G = \langle R(G)\rangle$. Then one of the following holds:

(1) $G = O^{p'}(\Delta)$.

(2) $\Delta = GL(V)$ and $G \cong Sp(V)$ acts naturally on V.

(3) $\Delta = GU(V)$ $|F| = q^2$, n is even, $G \cong Sp_n(q)$, and $V = F\otimes_{GF(q)}U$ where U is the natural GF(q)G-module.

(4) $|F| = 2$, $\Delta = GL(V)$ or $Sp(V)$, $G \cong O_n^{\epsilon}(2)$ or S_m, and V is the natural module for G.

(5) $|F| = 4$, $\Delta = GU(V)$, and one of the following holds:

(i) $G \cong O_n^{\epsilon}(2)$ or S_m and $V = F\otimes_{GF(2)}U$, where U is the natural GF(2)G-module.

(ii) $G \cong U_4(3)/Z_3$ and $n = 6$.

(iii) $G \cong S_n/E_{3^{n-1}}$ and G permutes a family $\mathcal{U} = (U_i : 1 \le i \le n)$ of nondegenerate points with V the direct sum of the points in $\mathcal{U}$.

Proof. By 6.5, G is transitive on $R = R(G)$. If $\Delta = GL(V)$ or $Sp(V)$, the result follows from a theorem of McLaughlin [28] and [29]. So assume $\Delta = GU(V)$. If p is odd let $X \in R$, $Y \in \theta(X)$, $K = \langle X,Y\rangle$, and $\langle z\rangle = Z(K)$. Then $K \cong SL_2(q)$, where $|F| = q^2$, and z is an involution with $K \trianglelefteq C_\Delta(z)$. Hence by [10] G is of Lie type over GF(q) or $q = 3$ and G is M_{11}. Then as

(*) $$\langle X,X_1\rangle \cong E_{q^2} \text{ or } SL_2(q) \text{ for all } X_1 \in R - \{X\},$$

we conclude $G/Z(G) \cong U_m(q)$ or $PSp_m(q)$. On the other hand if $p = 2$ then x^G is a set of odd transpositions of G, where $x \in X^{\#}$ and $X \in R$. Further X is a TI-set in G satisfing (*). We conclude from these observations and [8] and [10] that either $G/Z(G) \cong U_m(q)$ or $PSp_m(q)$, or $q = 2$ and x^G is a set of 3-transpositions of G.

Suppose $q > 2$. Then $G/Z(G) \cong U_m(q)$ or $PSp_m(q)$ and it is easy to check that G is generated by m members $(X_i : 1 \le i \le m)$ of R. So $V = [V,G] = \langle [V,X_i] : 1 \le i \le m\rangle$ is of dimension at most m. Thus $m \ge n$. As $U_{n+1}(q)$ is not involved in $U_n(q)$, $G = \Delta$ if $G/Z(G) \cong U_m(q)$. Similarly if $G/Z(G) \cong PSP_m(q)$ then $n = m$ is even. As GF(q) is a splitting field for G, $V = F\otimes_{GF(p)}U$ for some irreducible GF(q)G-module U. Then the members of R are root groups of transvections on U, so by Mclaughlin [28], (3) holds.

So take $q = 2$. Then x^G is a set of 3-transpositions of G, so G is described in [22]. Also $m([V,x]) = 2$, so G is described in 13.3 or 13.5 of [4], and reconciling these lists we find the lemma holds.

If $X = \langle R(X)\rangle \leq \Delta$ write $\mathcal{R}(X)$ for the set of subgroups $Y = \langle R(Y)\rangle$ of Δ with $R(X) \subsetneq R(Y) \neq R(\Delta)$.

(6.7) Assume $G = \langle R(G)\rangle$. Then $\mathcal{R}(G) = \{G\}$ precisely in the following cases:

(1) $\Delta = GL(V)$ and $G \cong Sp(V)$.

(2) $\Delta = GU(V)$, $|F| = q^2$, n is even, and $G \cong Sp_n(q)$.

(3) $|F| = 2$, $\Delta = Sp(V)$, and $G \cong O_n^{\epsilon}(2)$.

(4) $|F| = 4$, $\Delta = GU(V)$, $n = 6$, and $G \cong U_4(3)/Z_3$.

(5) $|F| = 4$, $\Delta = GU(V)$, and $G \cong S_n/E_{3^{n-1}}$ with $n \neq 2,3$, or 6.

Proof. Observe that if $G_0 \in \mathcal{R}(G)$ then G_0 is also irreducible on V, so G and G_0 are described in 6.6. Now to complete the proof observe that there exists no G_0 in 6.6 with G *L* G_0 *L* Δ precisely when G is listed in 6.7. Notice that in (5), n *h* 2 or 3 as then $R(G) = R(\Delta)$, while n *h* 6 as then G is contained in $U_4(3)$/Zy3. The groups of 6.6.5 are contained in the symplectic groups of 6.7.2.

(6.8) Let $\Delta = Sp(V)$ or $GU(V)$ and U a proper nondegenerate subspace of V with $\dim(U) > \dim(V/U)$. Then $\mathcal{R}(\langle N_\Delta(U)\rangle) = \langle N_\Delta(U)\rangle$.

Proof. As $\dim(U) > \dim(V/U) > 0$, $R(N_\Delta(U))$ is nonempty and indeed $D = \langle R(N_\Delta(U))\rangle = D_1 \times D_2$, where D_1 centralizes $U^\perp$, D_2 centralizes U, $D_1 \cong Sp(U)$ or $GU(U)$, and either $\dim(U^\perp) = 1$, $\Delta = GU(V)$, and $D_2 = 1$, or $D_2 \cong Sp(U^\perp)$ or $SU(U^\perp)$.

Suppose $G_0 \in \mathcal{R}(D)$ with D *L* G_0 *L* Δ. Claim G_0 is irreducible on V. For D_2 is irreducible on $U^\perp$ and D_1 is irreducible on U, so U and $U^\perp$ are the only proper nontrivial D-invariant subspaces of V. Thus as $D \neq G_0$, the remark follows.

So we may take $G = G_0$, and enlarging G if necessary, we may assume $\{G\} = \mathcal{R}(G)$. Thus G is described in 6.7. Now observe that $N_G(U)$ does not contain $R(N_\Delta(U)$ for any U and any G on the list of 6.7.

(6.9) Let V be the direct sum of $m > 1$ subspaces U_i of dimension $k \geq 2$. If Δ = Sp(V) or GU(V) assume the sum is orthogonal. Let $\Sigma = (U_i : 1 \leq i \leq m)$ and Γ = $SL(V) \cap \Delta$. Then one of the following holds:

(1) $N_\Gamma(\Sigma)$ is maximal in Γ.

(2) $k = 2$, $|F| = 2$ (or 4 if Δ = GU(V)), and $N_\Delta(\Sigma) \leq N_\Delta(G)$ where one of:

(a) Δ = GL(V) and $G \cong Sp_n(2)$.

(b) Δ = Sp(V) and $G \cong O_n^\epsilon(2)$ with $\epsilon = (-1)^m$.

(c) Δ = GU(V) and $G \cong S_n/E_{3^{n-1}}$ or $U_4(3)/Z_3$.

Proof. Let X_i be the group generated by members of $R = R(\Delta)$ acting on U_i and centralizing U_j for $j \neq i$. Let $X = \langle X_i : 1 \leq i \leq m\rangle$ and assume $N_\Gamma(\Sigma) < G_0 \leq \Gamma$. As $N_\Gamma(\Sigma)$ is irreducible on V, so is G_0, so we may take $G_0 = G$ and $X \leq H = \langle R(G)\rangle$. Then $X_i \leq H_j$ for some j, so $U_i \leq V_j$ and $\{U_i : U_i \leq V_j\}$ is a block of imprimitivity for $N_\Gamma(\Sigma)$ in its action on Σ. However $N_\Gamma(\Sigma)$ induces S_m on Σ and hence is primitive. Further as $G \neq N_\Gamma(\Sigma)$, these blocks are not of size 1, so we conclude H is irreducible on V. In particular H is described in 6.6. Then as $N_\Gamma(\Sigma) \leq G \leq N_\Gamma(H)$, we conclude (1) or (2) holds by checking when the stabilizer in Γ of a decomposition Σ is contained in $N_\Gamma(H)$ for some H in 6.7.

7. Some subgroups of alternating groups.

In this section p is a prime. Various facts about the alternating groups are collected.

(7.1) $((p^n)!)_p = p^{(p^n-1)/(p-1)}$.

Proof. $(p^n)!$ is the product of p^n factors and for each i, $1 \le i \le n$, exactly p^{n-i} of these factors are divisible by p^i. Therefore if $i < n$, exactly $p^{n-i} - p^{n-i-1}$ of the factors are divisible by p^i but not p^{i+1}. Thus if e is the power of p dividing $(p^n)!$, we conclude:

$$e = n + \sum_{i=1}^{n-1} (p^{n-i} - p^{n-i-1})i$$

$$= \sum_{j=0}^{n-1} p^j = (p^n - 1)/(p - 1).$$

(7.2) Let n be a positive integer and $n = \sum_{i=0}^{\infty} a_i p^i$ be the p-ary expansion of n. Then

$$(n!)_p = \prod_{i=0}^{\infty} (((p^i)!)_p)^{a_i}.$$

Proof. Induct on n. The result is trivial if $n = 1$. Let $k = \min\{i : a_i \neq 0\}$ and $m = n - a_k p^k$. Then $n! = (m!)(n!/m!)$ and by induction on n,

$$(m!)_p = \prod_{i=k+1}^{\infty} (((p^i)!)_p)^{a_i},$$

so it remains to show $(n!/m!)_p = (((p^k)!)_p)^{a_k}$. However $n \equiv a_k p^k \bmod p^{k+1}$, so $(n!/m!)_p = ((a_k p^k)!)_p$. Similarly $((rp^k)!/((r-1)p^k)!)_p = (p^k)!$ for $1 < r \le a_k$, so $((a_k p^k)!)_p = (((p^k)!)_p)^{a_k}$, as desired.

(7.3) Let n be a positive integer and $n = \sum_{i=0}^{\infty} a_i p^i$ be the p-ary expansion of n. Let Ω

$= \{1,\ldots,n\}$, G the symmetric group on Ω, and $(\Omega_{ij} : (i,j) \in \Delta)$ a partition of Ω by sets Ω_{ij} of order p^i, where Δ consists of all pairs (i,j) such that i is a nonnegative integer and $1 \leq j \leq a_i$. Let G_{ij} be the subgroup of G moving only points in Ω_{ij} and let $T_{ij} \in Syl_p(G_{ij})$. Then

(1) $T = \langle T_{ij} : (i,j) \in \Delta\rangle$ is the direct product of the subgroups T_{ij}, and is a Sylow p-subgroup of G.

(2) $(\Omega_{ij} : (i,j) \in \Delta)$ is the set of orbits of T on Ω.

(3) $Z(T_{ij}) = Z_{ij} \cong Z_p$ for $(i,j) \in \Delta$ with $i > 1$. Moreover Ω_{ij} is the set of points moved by Z_{ij}.

(4) Z(T) is the direct product of the groups Z_{ij}, $(i,j) \in \Delta$. Thus $m(Z(T)) = \sum_{i=1}^{\infty} a_i$.

Proof. Evidently T is the direct product of the groups T_{ij} and Z(T) is the direct product of the groups Z_{ij}. $G_{ij} \cong S_{p^i}$, so by 7.2, $|T| = |G|_p$. That is $T \in Syl_p(G)$. As $T_{ij} \in Syl_p(G_{ij})$ and $|\Omega_{ij}|$ is a power of p, T_{ij} is transitive on Ω_{ij}, and hence (2) holds. Further as T_{ij} is faithful and transitive on Ω_{ij}, Z_{ij} is semiregular on Ω_{ij}. Let z_{ij} be of order p in Z_{ij} and t_r, $1 \leq i \leq p^{i-1}$, the cycles of z_{ij} on Ω_{ij}. Z_{ij} centralizes each t_r and hence acts on the set of p points moved by t_r. So as Z_{ij} is semiregular on Ω_{ij}, $Z_{ij} = \langle z_{ij}\rangle$, completing the proof.

(7.4) Let L be the alternating group on $\Omega = \{1,\ldots,n\}$, $n \geq 5$, and G the alternating or symmetric group on Ω. Then

(1) Let $\Delta \subseteq \Omega$ with $|\Delta| \neq 0,n$, or $n/2$. Let $H = G(\Delta)$ be the subgroup of G acting on Δ. Then H is maximal in G.

(2) Let $\Sigma = (\Delta_i : 1 \leq i \leq m)$ be a partition of Ω with $|\Delta_i| = n/m$ for all i, $1 \leq i \leq m$, and with $m \neq 1$ or n. Let $K = G(\Sigma)$ be the subgroup of G preserving Σ. Then either K is maximal in G or $n = 2m = 6$ or 8 and $G = A_n$.

Proof. Assume M is a proper subgroup of G properly containing H or K. We first observe that M is transitive on Ω. This if clear if $K \leq M$ since K is transitive on Ω. Further H is transitive on Δ and on $\Omega - \Delta$, so if $H < M$ then M is transitive on Ω.

We next show M is primitive on Ω. For assume $\theta = \Gamma^M$ is a nontrivial system of imprimitivity for M. Suppose first $H < M$. Then $\{\alpha\cap\Delta : \alpha \in \theta\}$ is an H-invariant partition of Δ. So as H is primitive on Δ, either $\Delta \subseteq \alpha$ for some $\alpha \in \theta$ or $|\alpha \cap \Delta| \leq 1$ for each $\alpha \in \theta$. By symmetry the same holds for $\Omega - \Delta$, and so without loss $|\Delta| > |\Omega - \Delta|$. Hence $|\Delta| > |\Gamma|$, so $|\Delta \cap \alpha| \leq 1$ for each $\alpha \in \theta$. Then $|\theta| \geq |\Delta|$ so $|\Omega| = |\Gamma||\theta| \geq 2|\Delta| > |\Omega|$, a contradiction.

Suppose next $K < M$. Then $\theta' = \{\alpha \cap \beta : \alpha \in \Sigma, \beta \in \theta\}$ is a K-invariant partition of Ω. Pick $\alpha \in \Sigma$, $\beta \in \theta$ so that $\alpha \cap \beta$ is nonempty. The subgroup K_α of K fixing α pointwise is transitive on $\Omega - \alpha$ and fixes β, so $\beta \subseteq \alpha$. Also $K(\alpha)$ is primitive on α, so $|\beta| = 1$ or $\beta = \alpha$. As θ is nontrivial, the first case is out. As $K < M$, the second is out.

Thus M is primitive on Ω. In case (1) choose notation so that $|\Delta| > |\Omega - \Delta|$ and let $d = |\Delta|$. In case (2) let $d = |\Delta_1|$. If $d > 2$ there is X of order 3 in H or K moving exactly 3 points of Ω. Notice that in case (1) that as $n \geq 5$ and $|\Delta| > |\Omega - \Delta|$, $d > 2$. If $d = 2$ and $G = S_n$ there is a transposition X in K. Finally if $d = 2$ and $G = A_n$ at least there is a 4-group X in K moving exactly 4 points of Ω. As $X \leq M$ and M is primitive on Ω, Exercise 5.6.4 in [6] implies one of the following holds:

(a) $L \leq M$.

(b) Case (2) holds with $d = 3$ and $n = 6$.

(c) Case (2) holds with $d = 2$, $G = A_n$, and $n = 6$ or 8.

In (a), as M contains K or H, $M = G$, contrary to the choice of M. In (b), $K = N_G(P)$, $P \in Syl_3(G)$, so K is a Borel subgroup of $G \leq Aut(L_2(9))$, and hence is maximal in G. In (c) the conclusion of (2) holds.

(7.5) Let G be a group with $F^*(G) = L \cong A_n$, $n \geq 5$, and let A be a noncyclic elementary abelian p-subgroup of G. Then one of the following holds:

(1) $\langle \mathcal{N}(A) \rangle = G$.

(2) $n = 5$, $G = L$, $A \in Syl_2(G)$, and $\mathcal{N}(A) = \{N_G(A)\}$.

(3) $n = 2p$, $G = L$, $A \in Syl_p(G)$ has two orbits Δ_1 and Δ_2 on $\Omega = \{1,\dots,n\}$, and $\mathcal{N}(A) = \{K\}$, where K is the subgroup of G preserving the partition $\{\Delta_1, \Delta_2\}$ of Ω.

Proof. Let $(\Delta_i : 1 \leq i \leq m)$ be the orbits of A on Ω and choose notation so that $n_i \leq n_{i+1}$, where $n_i = |\Delta_i|$. If $m > 2$ then by 7.4, $G(\Delta_1)$ and $G(\Delta_2)$ are distinct maximal subgroups of G containing A, so (1) holds. Thus we may take $m \leq 2$.

Suppose $n_1 = 1$. Then $m = 2$ and $G(\Delta_1) \in \mathcal{N}$ by 7.4. Also $A \leq X \leq L$ with $X \cong$

$L_2(|A|)$ acting transitively in its representation of degree n on Ω. So (1) or (2) holds.

So $n_1 > 1$ and hence there is A_i of order p in A acting nontrivially on Δ_i. As A is abelian and transitive on Δ_i, A_i has n_i/p cycles $(\Delta_{ij} : 1 \le j \le n_i/p)$ on Δ_i. Now

$$\Sigma = (\Delta_{ij} : 1 \le i \le m,\ 1 \le j \le n_i/p)$$

is an A-invariant partition of Ω, so by 7.4, either $G(\Sigma) \in \mathcal{N}(A)$ or $p = 2$ and $n = 6$ or 8. In this last case $L \cong Sp_4(2)'$ or $L_4(2)$, so L has at least two A-invariant parabolics generating G, and we are done. Thus we may take $G(\Sigma) \in \mathcal{N}(A)$. But if $n_i \ne p$ we may make a different choice of A_i to obtain a different partition, and therefore (1) holds. This leaves us in the case $m = 2$, $n_1 = n_2 = p$, where (3) holds by 7.4.

Actually we haven't treated the case where $n = 6$ and $G \not\le S_6$, but this case is easy to handle by ad hoc methods.

(7.6) Let G be the alternating group on $\Omega = \{1,\ldots,n\}$ with $5 < n \le 9$. Then up to conjugacy in G there is a unique subgroup X of G isomorphic to A_m, $5 < m < 9$, and there are two subgroups X and Y isomorphic to A_5. $X = G_\Delta$, $\Delta = \{m+1,\ldots,n\}$ and Y moves 6 points of Ω.

Proof. This is because A_m has a unique quasiequivalence class of faithful representations of degree n unless $m = 5$, where it has two. Moreover the normalizer of the image of each such representation in S_n is not contained in A_n.

8. Large p-subgroups of finite groups.

In this section p is a prime, G is a finite group, and Q is a p-subgroup of G satisfying the following hypotheses:

(a) $\langle \mathcal{N}_G(Q,p')\rangle = B$ is cyclic.

(b) $C_G(Q) \leq Z(Q)B$.

(c) If $z \in \Omega_1(Z(Q))^{\#}$ and $Q \leq X = O^{p'}(X) \leq C_G(z)$ then either $E(X) = 1$ or $E(X)$ is quasisimple and $z \in O_p(E(X)) \leq Z(Q)$.

(d) One of the following holds:

(i) Q is of symplectic type and exponent p or 4 and $Q \trianglelefteq N_G(\Omega_1(Z(Q)))$.

(ii) If Y is a subgroup of G which is the direct product of $p^e > 1$ copies of a nonabelian simple group L, then $|Q| > |Aut(Y)|_p$.

Under these hypotheses the following result is established:

(8.1) Let $Q \leq M \leq G$. Then one of the following holds:

(1) $F^*(M) = O_p(M)$.

(2) $M \leq N_G(B_0)$ for some $1 \neq B_0 \leq B$.

(3) $M \leq N_G(Z_0)$ for some $1 \neq Z_0 \leq Z(Q)$.

(4) $F^*(M)$ is a nonabelian simple group.

(5) $Q \cong Q_8^2$, $F^*(M) \cong A_5 \times A_5$, and case (d.i) holds.

To prove the lemma take M to be a subgroup of G containing Q. If $O_{p'}(M) \neq 1$ then by (a), $O_{p'}(M) = B_0 \leq B$ so that 8.1.2 holds.

So take $O_{p'}(M) = 1$. Let $Z = \Omega_1(Z(Q))$. If $O_p(M) \neq 1$ then by (b) there is $1 \neq z \in Z \cap O_p(M)$. Let $X = O^{p'}(C_M(z))$. By (c) either $E(X) = 1$ or $E(X)$ is quasisimple and $z \in E(X)$. As $z \in O_p(M)$, $E(X) = E(M)$. Hence if $E(X) \neq 1$ then $1 \neq O_p(E(M)) = Z_0 \leq Z(Q)$, with the last containment a consequence of (c). But M normalizes Z_0 so 8.1.3 holds. If $E(X) = 1$ then 8.1.1 holds.

So we may assume $O_\infty(M) = 1$, and it remains to show one of the last two cases of lemma 8.1 holds. Let L be a component of M and $Y = \langle L^Q\rangle$. If $Y \neq F^*(M)$ then by (b) there is $z \in C_Z(Y)^{\#}$ and hence by (c), $Y = E(X) = L$ and $z \in L$. But then $O_\infty(M) \neq 1$, a contradiction. Hence $Y = F^*(M)$ and as it suffices to show $Y = L$, we may assume otherwise. We may also take $M = YQ$. In particular case (d.ii) cannot hold, leaving us in case (d.i).

In particular there is u of order p in $Q - N(L)$ and $Z = \langle z \rangle$ is of order p. Then $L^Q = \{L_1,...L_n\}$ and $z = z_1...z_n$ with z_i of order p in L_i. Choose notation so that $L = L_1$. Let $x \in C_L(z_1)$. Then $y = [u,x] \in LL^u \cap Q$ as $Q \trianglelefteq N_G(Z)$. Hence x is a p-element so $C_L(z_1)$ is a p-group. Also if p is odd then $1 \neq [u,y] \in LL^uL^{u^2}$, so $[u,y] = [Q,Q] = Z$ and hence as Z projects on each L_i we conclude $p = 3$. Further x is the projection of $[u,y]$ on L, so $x \in \langle z_1 \rangle$ and thus $|C_L(z_1)| = 3$. Therefore $L \cong L_2(4)$ or $L_3(2)$. In either case $\mathcal{U}_Y(Q,2)$ is noncyclic contrary to (b).

So $p = 2$. Similarly if x is not an involution then neither is $[u,y]$, contradicting $[Q,Q] = Z$. So $C_L(z_1) \cong E_{2^m}$. As Z is the unique subgroup of order 2 in $C_G(Q)$, z_1 is in the center of a Sylow 2-group of L, so $C_L(z_1) \in Syl_2(L)$. Hence $L \cong L_2(2^m)$. Also

$$E_{2^m} \cong V = \langle [u,x] : x \in C_L(z_1) \rangle \leq Q$$

so $[V,Q] \leq Z$ and hence $Y = L_1L_2$.

As $\Phi(Q) = Z \leq Y$ and $M = YQ$, M/Y is elementary abelian. So as $Out(L)$ is cyclic, $|Q:Q\cap Y| \leq 4$. Thus $Q = (Q\cap Y)\langle u \rangle$ or $(Q\cap Y)\langle u,v \rangle$ with v inducing an outer automorphism on L. In the latter case as $[V,v] \leq Z$, $m = 2$. Then $Q \cong Q_8^2$ and 8.1.5 holds. In the former case $V \leq Z(Q)$, a contradiction.

Notice the proof of lemma 8.1 is complete.

9. Centralizers of semisimple elements in groups of Lie type.

In this section L is a finite simple group of Lie type over GF(q), where q is a power of the prime r. G is a subgroup of the subgroup D of Aut(L) inducing inner-diagonal automorphisms on L. Let z be an element of prime order p, with $p \neq r$, and let T be a maximal torus of G containing z. Let $\tilde{L}$ be the univeral Chevalley group of type L.

(9.1) (1) $C_G(z)$ has normal subgroups Y, Y_0, and D_0 such that $Y = Y_0T$, $Y_0 = D_0Z$, $Z = C_Y(D_0) \leq T$, and D_0 is the central product of groups $D_i = O^{r'}(D_i)$ of Lie type over $GF(q^{e_i})$, $1 \leq i \leq k$.

(2) $|C_G(z)/Y|$ divides $|Z(\tilde{L})|_p$ and if p is odd or $Z(\tilde{L})$ has a cyclic Sylow p-group then $|C_G(z)/Y|$ divides p.

Proof. This is essentially lemma 2.5 in [34] or 2.9 in [35]. Let $\tilde{z}$ be a preimage of z in $\tilde{L}$. G is also a subgroup of Aut($\tilde{L}$), and subject to this identification, $Y = C_G(\tilde{z})$, $c = |C_G(z):Y|$ divides $|Z(\tilde{L})|_{p'}$ and if $Z(\tilde{L})$ has a cyclic Sylow p-group then c divides $(p,|Z(\tilde{L})|)$. Moreover if p is odd then $Z(\tilde{L})$ has a cyclic Sylow p-group.

In the remainder of this section adopt the notation established in 9.1. The following observation is trivial but useful:

(9.2) For each i, $1 \leq i \leq k$, one of the following holds:

(1) D_i is quasisimple.

(2) $q = q^{e_i} = 2$ and $D_i \cong SU_3(2)$, $U_3(2)$, $L_2(2)$, or $Sp_4(2)$.

(3) $q = q^{e_i} = 3$ and $D_i \cong SL_2(3)$ or $L_2(3)$.

The next lemma is well known:

(9.3) Let $j(L) = |Z(\tilde{L})|$. Then one of the following holds:

(1) $j(L)$ is a power of 2.

(2) $L \cong L_n^{\epsilon}(q)$ and $j(L) = (q-\epsilon,n)$.

(3) $L \cong E_6^{\epsilon}(q)$, $q \equiv \epsilon \bmod 4$, and $j(L) = 3$.

(9.4) There exists an automorphism α of G inducing the r-power map on T with αD generating the group of field automorphisms of $Aut(L)/D$.

Proof. There is a Chevalley group $\bar{G}$ and an endomorphism σ of $\bar{G}$ such that $O^{r'}(\bar{G}_{\sigma}) \leq G \leq \bar{G}_{\sigma}$, $L = O^{r'}(\bar{G}_{\sigma})$, and $Aut_{\bar{G}}(L) = D$. Let $\bar{T}$ be a σ-invariant maximal torus of $\bar{G}$ and β a Frobenius map on $\bar{G}$ inducing the r-power map on $\bar{T}$. By definition of T, $T = \bar{T}' \cap G$ for some σ-invariant conjugate $\bar{T}'$ of $\bar{T}$. Without loss of generality we may take $\bar{T} = \bar{T}'$. Now $[\beta,\sigma] \in \bar{T}$, so σ acts on the coset $\beta\bar{T}$, and hence by 2.7 in [44], σ fixes some $\alpha \in \beta\bar{T}$. Then α commutes with σ so α acts on $L = O^{r'}(\bar{G}_{\sigma})$ with αD a generator for the group of field automorphisms of $Aut(L)/D$. As each subgroup of D/L is normal in $\langle\alpha,D\rangle/L$, α also acts on G. As $\bar{T}$ is abelian, α induces the r-power map on $\bar{T}$ and hence also on T.

(9.5) Assume p is odd, $p^{1+2m} \cong Q \leq M \leq Aut(L)$, $G = D \cap M$, $M = GQ$, $Y = T$, $C_G(Q) = \langle z\rangle$, and $Q \trianglelefteq N_M(\langle z\rangle)$. Then either $G = M$ or $p = 3$, $L \cong D_4(q)$ or ${}^3D_4(q)$, and $|M:G| = 3$ with some element of Q inducing an automorphism on L in the coset of a graph or graph-field automorphism.

Proof. Either the lemma holds or for some $x \in Q$, $xD = \beta D$ for some nontrivial field automorphism β of L of order p. Of course we may assume the latter. By 9.3, $xD = \alpha D$ for some $\alpha \in M$ inducing the r^e-power map on T, where $q = r^{pe}$.

Let T_0 be the subgroup of T generated by the elements of T of order p. Then $[x,T] \leq Q \cap T \leq T_0$ (as Q is of exponent p) and $[Q \cap T,x] \leq \langle z\rangle$ as $Q \trianglelefteq C_G(z) \geq T$ and Q is extraspecial.

Suppose $(j(L),p) = 1$. Then by 9.1, $Y = C_G(z)$ and by hypothesis $T = Y$. Hence $x \in \alpha T$, so that x induces the r^e-power map on T. Therefore x centralizes T_0. Hence $m(Q \cap T) + 1 = m((Q \cap T)\langle x\rangle) \leq m(Q) = m+1$, so $m(QT/T) \geq m+1 \geq 2$. But as $T = C_G(z)$ and $|QG:G| = p$, $m(QT/T) = 1$, a contradiction.

So p divides $j(L)$. Therefore by 9.3, $L \cong L_n^{\epsilon}(q)$ or $E_6^{\epsilon}(q)$ and p divides $(q-\epsilon,n)$ or $(q-\epsilon,3)$, respectively. In particular T is a subgroup of index d dividing $j(L)$ of the direct product of k cyclic groups of order $q - \epsilon$, and the Weyl group $W = N_G(T)/T$ of T is isomorphic to $W(A_{n-1}) \cong S_n$ or $W(E_6) \cong O_6^-(2)$, respectively.

As Q is extraspecial and $C_G(Q) = \langle z\rangle$, $\langle z\rangle = Z(P)$ for $P \in Syl_p(M)$ with $Q \leq P$. As $Y = T$, $|P\cap G:P\cap T| = p$ by 9.1.2. But $|O_6^-(2)|_3 = 3^4$, so L is not $E_6^\epsilon(q)$. Thus L is $L_n^\epsilon(q)$ and a similar argument shows $n = p$.

Next either $x \in \alpha T$ and $|T/C_T(x)| = (q-\epsilon)^k p^a/(r^e-\epsilon)^k$, $a \leq 1$, or $x = \alpha y$ for some $y \in N_G(T) - T$ and y is faithful on $O_{p'}(T/C_T(\alpha))$. However $[x,T] \leq T_0$, so $|T:C_T(x)|$ is a power of p. It follows that $(q-\epsilon)/(r^e-\epsilon)$ is a power of p. Therefore $p = 3$ and $L \cong U_3(8)$.

Observe $|Q| \leq |T_0||N_G(T)/T|_3 = 3^4$. Therefore $Q \cong 3^{1+2}$. Further as $\Omega_1(\langle\alpha\rangle T) = \langle\alpha\rangle T_0 \cong E_{27}$ we conclude $Q \not\leq \langle\alpha\rangle T$. But then $T_0 \leq [T,Q] \leq Q$, so $Q = T_0\langle x\rangle$ and $x = \alpha y$ for $y \in N_G(T) - T$. However now QT/T is not invariant under an involution in $N_G(T)$, contradicting $Q \trianglelefteq N_M(\langle z\rangle)$.

Assume in the remainder of this section that r is odd and L is not $L_2(q)$ or ${}^2G_2(q)$. I recall some facts from [3] and [10]. A *fundamental subgroup* of G is a subgroup of the form $K = \langle X,X^-\rangle$, where X is a long root group of L or L is $G_2(q)$ or ${}^3D_4(q)$ and X is a short root group of L, and in any event X^- is the negative of X. Then $K \cong SL_2(q)$, unless $L \cong {}^3D_4(q)$ and $K \cong SL_2(q^3)$. Moreover L is transitive on its fundamental subgroups unless L is $G_2(q)$ or ${}^3D_4(q)$, where there are two orbits. In particular K has quaternion Sylow 2-subgroups and a unique involution z(K). For $X \leq G$, denote by R(X) the set of Sylow 2-subgroups of fundamental subgroups of G contained in X and if $U \leq R \in R(G)$ with $|U| > 2$, write K(U) for the unique fundamental subgroup of G containing U.

Let $S \in Syl_2(G)$. Define Fun(S) to be the set of fundamental subgroups K of G such that $K\cap S \in Syl_2(K)$. We find in 1.3 of [3] that:

(9.6) Distinct members of Fun(S) commute. Indeed for $R,Q \in R(G)$, $[R,Q] = 1$ if and only if $[K(R),K(Q)] = 1$.

For $X \leq G$, define U(X) to be the set of cyclic subgroups of X which are of index 2 in some member of R(G). Define $\eta(S)$ to be the collection of subsets η of U(S) of maximal order subject to $K(U) \neq K(V)$ for distinct U and V in η. Define $\eta(G)$ to consist of those α such that for some $S \in Syl_2(G)$, $\eta \in \eta(S)$ and η is

S-invariant. By 1.1, 1.2, and 1.4 in [3]:

(9.7) (1) There exists $\alpha \in \eta(G)$.

(2) $\langle\eta\rangle$ is abelian and $\langle\eta\rangle \cap U(G) = \eta$.

(3) Let $\alpha \subseteq S \in Syl_2(G)$. Then either α is weakly closed in S or $K \cap S \in R(G)$ for each fundamental subgroup K of G with $|K \cap S| > 2$.

(4) G is transitive on $\eta(G)$.

Define the *Thompson group* of G to be $\mu(G) = \langle R(N_G(\eta))\rangle/\langle\eta\rangle O(\langle R(N_G(\eta))\rangle)$, where $\eta \in \eta(G)$. By 9.7 the Thompson group is independent of the choice of η.

We next use the theory in [3] and [10] to determine the overgroups of S for certain choices of L.

(9.8) Let G be $G_2(3)$, $L_4^{\epsilon}(3)$, or $P\Omega_8^{+}(3)$. Then the maximal overgroups of a Sylow 2-subgroup S of G are as described in the tables of the introduction.

Proof. This was essentially done in [15]; a few more details are supplied here. First let $\langle z\rangle = Z(S)$, $H = C_G(z)$, and $Q = F^*(H)$. Observe Q is an extraspecial 2-group and $Q = Q_1 \cdots Q_k$, where $Q_i \in R(G)$, $K_i = K(Q_i)$, and $\{K_i : 1 \le i \le k\} = Fun(S)$.

Notice hypotheses L1-L3 of section 17 are satisfied by Q in G. Claim $\mathcal{V}_G(Q,2') = 1$, so that L4 holds too. This follows from 17.2, 17.3, and 17.5 together with the observation that z is not weakly closed in Q with respect to G unless G is $L_4(3)$, where at least $\mathcal{V}_{C(x)}(C_Q(x),2') = 1$ for x an involution in $Q - \langle z\rangle$.

Let $\alpha \in \eta(G) \cap S$ and $W = \langle\alpha\rangle$. By 9.7.3 and [10] (Particularly Table 7.18 of [10], which tells us when $\mu(G)$ is abelian and hence when the exceptional case of 9.7.3 arrises.), either W is weakly closed in S or G is $L_4(3)$ and $W \cong Z_4 \times Z_2 \le Q$. Moreover in the former case $N_G(W) = \langle R(N_G(W))\rangle$ and $\mu(G) \cong D_{12}$, S_4, or $Wyl(D_4)$ for $G \cong G_2(3)$, $U_4(3)$, or $P\Omega_8^{+}(3)$ respectively. If G is $L_4(3)$ then there are just two normal E_{16}-subgroups V_1 and V_2 of S, V_1 is fused to V_2 in Aut(G), and $N_G(V_i)/V_i \cong S_5$.

Let $S \le M < G$ with $M \not\le H$. As $\mathcal{V}_G(Q,2') = 1$, $O(M) = 1$. For $R \in R(M)$ let $I(R) = K(R) \cap M$. Then $I(R) = R$ or $K(R)$, and $\Omega = \{I(R) : R \in R(M)\}$ satisfies Hypothesis Ω of [10] in M. Thus we can appeal to Theorems 1 through 8 of [10] to determine the structure of $\langle\Omega\rangle$.

Suppose first $O_2(M) \ne 1$. Then by 17.4, $z \in F^*(M) = O_2(M)$ and then by Theorem 2 in [10] and as S is Sylow in M, one of the following holds:

(a) G is not $L_4(3)$ and $M = N_G(W)$.

(b) G is $G_2(3)$ and $M \cong L_3(2)/E_8$.

(c) G is $U_4(3)$ and $M \cong A_6/E_{16}$.

(d) G is $P\Omega_8^+(3)$ and M is a parabolic of A_8/E_{64}.

(e) G is $L_4(3)$ and $M = N_G(V_i)$.

On the other hand if $O_2(M) = 1$ then by 17.4, $F^*(M)$ is a nonabelian simple group. Then as $M \leq G$ and $S \in Syl_2(M)$, we conclude from Theorem 1 in [10] that M is isomorphic to one of the subgroups of G listed in the tables in the introduction.

In particular as a consequece of the last two paragraphs and the structure of the possible subgroups listed there, we find:

(9.8.1) If G is not $L_4(3)$ then $M = \langle H\cap M, N_G(W)\rangle$. If G is $L_4(3)$ then $M = \langle H\cap M, N_G(V_i)\rangle$, for i = 1 or 2.

Now let G be $G_2(3)$. Then H has 4 subgroups X_1, X_2, Y_1 and Y_2 of index 3 containing S with $O^2(X_i) = K_i$ and $O^2(Y_i) \cong Z_3/Q^2$. It is well known that G contains a subgroup $G_3 \cong G_2(2)$. Then $H\cap G_3 \cong Z_2/(Q_8 * SL_2(3))$ so $H\cap G_3 = X_1$ or X_2, say X_1. Further $G_4 = G_3^a = \langle X_2, N_G(W)\rangle$. Next $\Omega_1(O_2(N_G(W)))) = V \cong E_8$ and an argument in the proof of 23.2 shows $N_G(V)/V \cong L_3(2)$. So $G_1 = N_G(V) = \langle Y_1, N_G(W)\rangle$, up to choice of notation. Finally let $G_0 = \langle Y_2, N_G(W)\rangle$. If $G_0 \neq G$ then by remarks above, $G_0 \cong G_2(2)$ or $L_3(2)/E_8$, and as $O^2(H\cap G_0) = O^2(Y_2) \cong Z_3/Q^2$ it must be the latter. This is impossible as V is the unique normal E_8-subgroup of $N_G(W)$. It follows from 9.8.1 that $\mathcal{N}(S) = \{G_1, H, G_3, G_4\}$, completing the analysis of $G_2(3)$.

Let G be $U_4(3)$. Here $\mathcal{N}_H(S) = \{H_1, H_2\}$ and $H_2 = (H_1)^a$ for suitable $a \in N_{Aut(G)}(S)$. Further from [10] there is $A_6/E_{16} \cong G_1 \leq G$. Thus by 9.8.1 and the enumeration of the possible isomorphism types of overgroups of S above, H, $G_1 = \langle H_1, N_G(W)\rangle$ and $G_3 = (G_1)^a$ are the maximal overgroups of S in G.

The arguments above and the discussion on page 1679 of [15] suffice to handle the case $G \cong P\Omega_8^+(3)$. This leaves $G \cong L_4(3)$. The stabilizer G_1 in G of a nonsingular point of the 6-dimensional orthogonal space preserved by G is isomorphic to $P\Omega_5(3)$. Also there is an outer automorphism a of G acting on S with $G_2 = (G_1)^a \neq G_1$. There is an S_5/E_{16}-subgroup of G_i which we may take to be $N_G(V_i)$. Thus by 9.8.1, $G_i = \langle H, N_G(V_i)\rangle$, i = 1,2, are the two members of $\mathcal{N}(S)$.

10. Centralizers of unipotent elements in groups of Lie type.

In this section L is a finite simple group of Lie type over GF(q), where $q = p^e$ and p is a prime. G is a subgroup of Aut(L) containing L, $T \in Syl_p(G)$, and $z \in Z(T)^{\#}$.

(10.1) $Z(T) \leq L$ and one of the following holds:

(1) Z(T) is contained in a long root group of L.

(2) p = 2 and $L \cong Sp_n(q)$ or $F_4(q)$.

(3) p = 3 and $L \cong G_2(q)$.

Proof. It is well known that, except in the exceptional cases of 10.1.2, T acts on a long root group of L. Hence 10.1 follows from the following result:

(10.2) Assume L is not $Sp_n(2^e)$, $F_4(2^e)$, or $G_2(3^e)$, let R be a long root group of G, $M = N_G(R)$, and $Q = O_p(M \cap L)$. Then one of the following holds:

(1) $Q = F^*(M)$, Q is special, and $R = Z(Q)$.

(2) $L \cong {}^2F_4(2^e)$, $e > 1$, $R = Z(Q)$, and $Q = F^*(M)$ is of class 3.

(3) $L \cong {}^2G_2(3^e)$, $e > 1$, and $R = Q = F^*(M)$ is of class 3.

(4) $L \cong L_3(2)$ or $L_3(4)$ and some element of T induces a graph automorphism on L.

(5) $G \cong Aut(L_4(2))$.

(6) $L \cong L_2(q)$, ${}^2F_4(2)'$, $G_2(2)' \cong U_3(3)$, or ${}^2G_2(3)' \cong L_2(8)$.

Proof. Let $X = M \cap L$. Again it is well known that $Q = F^*(X)$, $C_M(Q) \leq Q$, and that unless $L \cong L_2(q)$, ${}^2F_4(q)'$, or ${}^2G_2(q)'$, Q is special with center R. So it remains to show T/Q is faithful on X/Q, unless one of the exceptional cases of 10.2 holds. So assume $t \in T-Q$ with $[t,X] \leq Q$. If t induces a field automorphism on L then t induces a field automorphism on X/Q which turns out to be faithful. Thus $p \leq 3$ and t induces a graph or graph-field automorphism on L. Here t induces a graph or graph-field automorphism on X/Q, so that t centralizes X/Q only when $L \cong L_3(2)$, $L_3(4)$, or $L_4(2)$ and t induces a graph automorphism.

(10.3) Assume the hypotheses and notation of 10.2 with $Q = F^*(M)$ special. Then one of the following holds:

(1) M is irreducible on Q/R.

(2) $L \cong L_n(q)$ and $Q/R = (Q_1/R) \times (Q_2/R)$ with Q_i/R the irreducibles of M on Q/R. Q_i is elementary abelian.

(3) $L \cong L_3(2), L_4(2)$, or $L_3(4)$.

Proof. Once again the remarks are well known.

(10.4) Let $G \in \mathrm{Chev}(p) \cup \{^2F_4(2), G_2(2)\}$ with G not A_6 if $p = 2$. Let Z be a long root group of G, $H = N_G(Z)$, $Q = O_p(H)$, Ω the set of conjugates A of Z such that $\langle A,Z\rangle$ is not a p-group and $A \in \Omega$. Then

(1) Q is transitive on Ω unless $p = 3$ and $G \cong {}^2G_2(3)'$ or $p = 2$ and $G \cong G_2(2)'$ or the Tits group. In the last case at least H is transitive on Ω.

(2) $G = \langle Q,A\rangle = \langle Q,a\rangle$ for each $a \in A^\#$ and $A \in \Omega$.

Proof. If $G \cong {}^2G_2(3)' \cong L_2(8)$ then (1) holds vacuously while $\mathcal{N}(Q) = \{N_G(Q)\}$, so (2) holds. Thus we may assume G is not ${}^2G_2(3)'$. First a sketch of the proof of (1) when G is not the Tits group or $G_2(2)'$. By Exercise 14.6 in [6], the rank r of G as a permutation group on the cosets of H is the same as that of the Weyl group W of G on the cosets of the stabilizer W_H in W of H. The latter is easily caluculated. (cf. Exercise 14.6 in [6]) Moreover it is also easy to find r conjugates of Z in distinct H-orbits. (cf. 12.1 of [12]) In particular it turns out there is a unique orbit A^H with $K = \langle A,Z\rangle$ not a p-group. Indeed if Z is the root group of the root α then A can be taken to be the root group of $-\alpha$. Then an easy calculation using the Chevalley commutator relations (cf. 4.2 in [13]) shows A is normalized by a Levi factor of H. As Levi factors are complements to Q in H, Q is transitive on Ω.

So let $G = G_0'$ with $G_0 \cong {}^2F_4(2)$ or $G_2(2)$. In these cases $|N_{G_0}(A):N_H(A)| = |G_0:G| = 2$, so at least H is transitive on Ω. Further if G is the Tits group then even $O_2(N_{G_0}(Z)) \le QN(K)$ so Q is transitive on K^H.

So (1) is established. To prove (2) we use (1) and argue as in 4.5 of [13]. Namely let $X = \langle Q,a\rangle$; we wish to show $H \le N_G(X)$. Now $K = \langle Z,a\rangle$ so $X = \langle Q,K\rangle = \langle Q,A\rangle$. Also $Q \trianglelefteq$

H so $N_H(K) \leq N_G(X)$. Thus $H = QN_H(K) \leq N_G(X)$ unless G is $G_2(3)'$. There at least $O^2(H) \leq N_G(X)$ so arguing as in the proof of 9.8, X = G.

We may assume G is not $G_2(2)'$, and replacing G by G_0, we may assume G is not the Tits group. Thus as $H \leq N_G(X) = M$, M is a parabolic of G. As $F^*(M) = O_p(M)$ and Z is in the center of a Sylow p-subgroup of M, $Z \leq O_p(M)$. This is impossible as $K \leq M$ but $Z \not\leq O_p(K)$.

(10.5) Assume the hypthesis and notation of 10.4. Then $\mathcal{N}(Q)$ is the set of maximal parabolics of G containing Q.

Proof. Let $M \in \mathcal{N}(Q)$. If $O_p(M) \neq 1$ then the result holds by the Borel-Tits Theorem. Thus we may take $O_p(M) = 1$, so by Baer-Suzuki there exists $x \in M$ with $\langle Z,Z^x \rangle$ not a p-group. But now G = M by 10.4.

11. 3-structure in Sz and Ly.

In this section G is Sz or Ly, $T \in Syl_3(G)$, and $Z = Z(T)$. We'll see that the 3-local structure and T-overgroups of Sz and Ly are markedly similar. At various points we'll need to know facts about the 3-locals and overgroups of certain proper sections of G. These facts will be formally derived in later sections; appeals to such facts are made here in an inductive setting. The structure of normalizers of subgroups of order 3 is assumed and can be found in [23]; for convenience of the reader this structure is listed in lemma 11.1:

(11.1) (1) T is of order 3^7.

(2) $Z \cong E_9$.

(3) $N_G(T)/T \cong SD_{16} \times Z_2\epsilon$, where $\epsilon = 0$ or 1 for $G \cong Sz$ or Ly, respectively. $N_G(T)$ has two orbits of length 2 on the subgroups of Z of order 3. Let Z_1 and Z_2 be representatives for these orbits.

(4) $N_G(Z_1) \cong Z_2/U_4(3)/Z_3$ or $Aut(Mc)/Z_3$ for $G \cong Sz$ or Ly, respectively. Also $Z_1 \leq E(N_G(Z_1))$.

(5) $F^*(N_G(Z_2)) = O_3(N_G(Z_2))$ is special of order 3^6 with center Z. $N_G(Z_2)/O_3(N_G(Z_2)) \cong SL_2(3)*D_8$ or $Z_2/(Z_4*SL_2(3))$ for $G \cong Sz$ or Ly, respectively.

(6) G has c conjugacy classes of subgroups of order 3 with representatives $(Z_i : 1 \leq i \leq c)$; c = 2 or 3 for $G \cong Sz$ or Ly, respectively. $N_G(Z_3) \cong (Z_2/E_9) \times A_6$ with $O_3(N_G(Z_3)) = D$ a TI-set in G and $N_G(D) \cong Q_8/(E_9 \times A_6)$.

In the remainder of this section let $Q = O_3(N_G(Z_2))$.

(11.2) (1) $C_G(Q) = Z$.

(2) Q is weakly closed in T with respect to G.

Proof. By 11.1.5, $Q = F^*(N_G(Z_2))$, so (1) holds. Part (2) follows from 11.1 and 2.3.

(11.3) Let $Q \leq X \leq N_G(Z_1)$ and $Y = \langle Q^X \rangle$. Then one of the following holds:

(1) $Y = Q$.

(2) $Y = E(N_G(Z_1))$.

(3) $Y \cong A_6/E_35$ and Y/Z_1 is generated by the 3-elements in a maximal parabolic of $E(N_G(Z_1))/Z_1 \cong U_4(3)$ or Mc.

(4) $G \cong Ly$ and $Y \cong U_4(3)/Z_3$.

Proof. As $Q/Z_1 = O_3(N_G(Z))/Z_1 \cong 3^{1+4}$, these remarks follow from 11.1.4 and the nature of the 3^{1+4}-overgroups of $U_4(3)$ or Mc. (eg. 10.5 and 26.2.) Remember this last appeal is in an inductive setting.

(11.4) $J(T) = A \cong E_35$, $N_G(A)/A \cong M_{11} \times Z_2\epsilon$, (this is the ϵ of Lemma 11.1.3) $|Z_1^G \cap A| = 11$, $|Z_2^G \cap A| = 110$, and $Z_3^G \cap A$ is empty.

Proof. From the structure of $N_G(Z_1)$, $J(T) = A \cong E_35$ and setting $M = N_G(A)$, $N_M(Z_1)/A \cong M_{10} \times Z_2\epsilon$ has orbits of length 1,10,20,90 on the set Δ of subgroups of A of order 3, with the orbits of length 1 and 10 in $\Gamma = Z_1^G \cap A$. As A is weakly closed in T, M is transitive on Γ. The orbit of length 20 is Z_2^M, so $|\Gamma| = 11$ or 101, and as 101 is a prime which does not divide the order of $GL_5(3)$, $|\Gamma| = 11$. As $m_3(C_G(Z_3)) = 4$, $Z_3^G \cap A$ is empty, so $|Z_2^G \cap A| = 110$. Now M^Γ is 2-transitive of degree 11 with point stabilizer M_{10}, so $M^\Gamma = M/O_{3,2}(M) = M_{11}$. As $A = O_3(M)$ and $|O_{3,2}(M)|_2 = 2^\epsilon$ the lemma follows.

(11.5) $\mathscr{V}_G(Q,3') = 1$.

Proof. This is because $\mathscr{V}_{C(X)}(Q,3') = 1$ for each X of order 3 in Z. (cf. 2.2)

(11.6) Let $Q \le H < G$. Then either:

(1) $F^*(H) = O_3(H)$, or

(2) There is $X \in Z_1^{N(T)}$ with $X \trianglelefteq H$ and $E(C_H(X)) \cong U_4(3)/Z_3$ or Mc/Z_3.

Proof. By 2.1, 11.1, 11.2, and 11.3, (G,Q) satisfies the hypotheses of section 8. Hence 8.1, 11.1, and 11.3 reduce us to the case where $F^*(H) = L$ is a nonabelian simple group. By 11.2.2 we may take $S = T \cap H \in Syl_3(H)$. Thus $S = Q$ or T. So $\Omega_1(S)$ is nonabelian and $Z = Z(S) \cong E_9$. It follows from the structure of Sylow 3-groups of sporadic groups (cf. [23]) that if L is sporadic then $L \cong Sz$ or Ly. But then $|Sz| < |Ly|$ and $|Sz|$ does not divide $|G|$.

Suppose $L \cong A_n$. As $|S| \geq 3^6$, $n \geq 15$. As $m_5(L) \leq 5$, $n < 18$. But now $|Z(S)| = 27$, a contradiction.

So $L \in Chev(r)$ for some prime r. Without loss $H = LQ$. From the structure of $Aut(L)$, $Z = Z(S) \leq L$.

Assume $r = 3$. If $L \cong {}^2G_2(q)$ then $|Z(S)|$ is an odd power of 3, a contradiction. If $L \cong G_2(q)$ then as S is of order at most 3^7, $q = 3$. Then S is of order 3^6 so $S = Q$. Further there is X of order 3 in Z with $O^{3'}(C_H(X)) \cong SL_2(3)/(E_9 \times 3^{1+2})$, contradicting 11.1 and 11.3.

So L is neither of these groups and as $Z = Z(S)$, Z is contained in a long root group R of G by 10.1. So $Q \leq O_3((C_H(Z_2)) \leq O_3(C_H(R)) \leq L$ and hence $H = LQ = L$. So $R = Z(S) = Z$ and therefore $q = 9$. As $|S| \leq 3^7$ and S is nonabelian, $L \cong L_3^\epsilon(9)$. But then $Z^\#$ is fused in L, contrary to 11.1.

So $r \neq 3$. Suppose Q is not normal in $N_H(X)$ for some X of order 3 in Z. Then by 11.1 we may take $X = Z_1$ and by 11.3, $\langle Q^{(N(X) \cap H)}\rangle \cong Mc/Z_3$, $U_4(3)/Z_3$, or A_6/E_{3^5}. But by 9.1, $N_L(X)$ has no such subgroup. Thus $\Gamma_{1,Z}(H) \leq N_H(Q) < H$, so L is one of the groups listed in Theorem 1 of [34]. As $|S| = 3^6$ or 3^7 we conclude $H \cong Sp_{10}(2)$, $U_6(2)$, $PGU_6(2)$, $Z_3/\Omega_8^+(2)$, $\Omega_{10}^-(2)$, $F_4(2)$, or $F_4(4)$. But in each case $|Z(S)| = 3$ or 27, a contradiction.

This completes the proof of 11.6.

(11.7) $\mathcal{N}(Q) = \mathcal{N}(T)^{N(Q)}$ and $\mathcal{N}(T) = \{M_i : 1 \leq i \leq 4\}$ where:

(1) $M_1 = N_G(J(T))$.

(2) $M_2 = N_G(Z)$.

(3) $M_3 = N_G(Z_1)$ and $M_4 = (M_3)^g$ for $g \in N_G(T) - N_G(Z_1)$.

Proof. Let $M \in \mathcal{N}(Q)$. By 11.4 and 11.6 we may assume $F^*(M) = O_3(M) = P$ and it remains to show either Q or some $N(Q)$-conjugate of $A = J(T)$ is normal in M. Let $Q \leq S \in Syl_3(M)$. By 11.6.2, $Q \trianglelefteq S$, so conjugating in $N_G(Q)$ we may assume $S = T \cap M$. In particular $Z = Z(S)$ and $S = Q$ or T. Let $V = \langle Z^M \rangle$. $Z \leq C_M(P) = Z(P)$, so V is abelian and as $M \not\leq N(Q)$, $Z \neq V$.

Claim $Q \trianglelefteq N_M(X)$ for each X of order 3 in Z. For if not by 11.1 we may assume Y

$= \langle Q^{(M\cap N(Z_1))}\rangle$ is not Q, so that Y is described in 11.3.3. (In 11.3.2, $O_3(Y) = Z_1$ whereas $V \leq O_3(Y)$.) In particular $A = \langle Z^Y\rangle \leq \langle Z^M\rangle = V$, so $V \leq C_G(A) = A$ by 11.4, and thus $M \leq N_G(V) = N_G(A)$, contrary to the hypothesis that A is not normal in M.

It follows from the claim and 11.2.2 that $Z = C_V(Q)$ is a TI-set in M. Let $M^* = M/C_M(V)$ and $\Delta = Z^M$. Then for $D \in \Delta - \{Z\}$, $m(ZD) = 4$, so as $V \neq A$ and $m(V) = 5$, we conclude $V = ZD$. Claim Q^* is semiregular on $\Delta - \{Z\}$. For if $x^* \in Q^*$ fixes $D \in \Delta - \{Z\}$ then x centralizes $ZD = V$, so $x^* = 1$. Notice also $8\,|\,|\Delta| \leq |V^\#| = 80$, so $|\Delta| \leq 10$. Hence as all orbits of Q^* on $\Delta - \{Z\}$ are of length 9, Q^* is regular on $\Delta - \{Z\}$ and Δ partitions V.

Now M^* is 2-transitive on Δ with $E_9 \cong Q^* \trianglelefteq N_{M^*}(Z)$ and Q^* regular on $\Delta - \{Z\}$. So setting $L = \langle Q^M\rangle$, $L^\Delta = L_2(9)$ on 10 letters. As $C_V(x^*) = Z$ for each $x^* \in Q^{*\#}$, $L \cong SL_2(9)$ acts naturally on V. But then $Z^\#$ is fused in L, contrary to 11.1.

12. 2-structure in certain sporadic groups.

In this section G is one of the seven sporadic groups Mc, Ly, Ru, ON, Co_3, M(22), or M(23). T is a Sylow 2-subgroup of G and z is an involution in the center of T. Let Q = T unless G is Ru or M(22), where $Q = O_2(C_G(z))$. From [23]:

(12.1) (1) $Z(Q) = \langle z \rangle$ unless G is M(22) or M(23).

(2) If G is M(22) then $Q \cong Z_2 \times D_8^4$.

(3) If G is M(23) then $Z(Q) \cong E_8$.

(12.2) Let $z' \in Z(Q)^{\#}$. Then the isomorphism type of $C_G(z')$ is listed in Table 12.2.

TABLE 12.2

G	Mc	Ly	Ru	M(22)
$C_G(z')$	A_8/Z_2	A_{11}/Z_2	$S_5/2^{11}$	$U_6(2)/Z_2$, $O_6^-(2)/D_8^4 \times Z_2$

G	Co_3	ON
$C_G(z')$	$Sp_6(2)/Z_2$	$Z_2/L_3(4)/Z_4$

G	M(23)
$C_G(z')$	$M(22)/Z_2$, $Z_2/U_6(2)/E_4$, $Z_2/(Z_3 \times U_4(2))/(E_4 \times D_8^4)$

(12.3) Let $B = \langle \mathcal{N}_G(Q,2') \rangle$. Then either $B = 1$ or $G \cong Ly$, $B \cong Z_3$, and $N_G(B) \cong Aut(Mc)/Z_3$.

Proof. Let $E_4 \cong E \trianglelefteq T$ with $E \leq Z(Q)$ if possible. Then each member of E is fused into $Z(Q)$ under G; this is because $|C_G(t)|_2 < |G|_2/2$ if G is not $M(22)$ and t is a non-2-central involution of G. In particular $C_G(e)$ is described in 12.2 for each e in $E^{\#}$. Let $B_e = \langle \mathcal{N}_{C_G(e)}(C_Q(e),2')\rangle$. From the nature of the $C_Q(e)$-signalizers of $C_G(e)$, either $B_e = 1$ or G is Ly and $B_e = B \cong Z_3$ is independant of e and $C_G(e)\cap N(B_e) \cong S_8/Z_6$. In this last case from the list of normalizers of subgroups of order 3 in [23], $N_G(B) \cong Aut(Mc)/Z_3$. (As usual there is an inductive appeal to the nature of these signalizers when $C(e)$ is a covering of $M(22)$.) Hence 2.2 completes the proof.

(12.4) Q is weakly closed in T with respect to G.

Proof. If $Q = T$ the remark is clear; in the remaining cases appeal to 2.3.

(12.5) Let $Q \leq M < G$. Then one of the following holds:

(1) $F^*(M) = O_2(M)$.

(2) G is Ly and $M \leq N_G(B)$, where B is defined in 12.3.

(3) $M \leq C_G(z')$ for some $z' \in Z(Q)^{\#}$.

(4) $F^*(M)$ is a nonabelian simple group.

Proof. Observe first that (G,Q) satisfies the hypotheses of section 8. Namely 12.3 establishes hypothesis (a) of section 8. Hypotheses (b) and (c) follow from the structure of the centralizers listed in 12.2 and the nature of the Q-overgroups in those centralizers. Finally 2.1 establishes hypothesis (d).

Next comes an appeal to 8.1 to conclude one of the first four possiblities listed in that lemma holds. Of course three of these possibilities correspond to cases of 12.5, so it remains to consider possibility three of 8.1. By 12.1, if $1 \neq Z_0 \leq \Omega_1(Z(Q))$ then either $Z_0 = Z(Q)$ is of order 2 and hence (3) holds, or G is $M(22)$ or $M(23)$. Even in these latter cases $N_G(Z_0) \leq C_G(z')$ for some $z' \in Z(Q)^{\#}$, so that (3) holds. (cf. 12.13)

(12.6) Let G be Ly or Mc. Then

(1) $\mathcal{A}(T) = \{A_1,A_2\}$ with $A_i \cong E_{16}$ and $J(T)$ isomorphic to a Sylow 2-group of $L_3(4)$.

(2) If G is Mc then A_1 is normal in T but fused to A_2 in $Aut(G)$. If G is Ly then A_1 is fused to A_2 in T.

(3) $BA_i = C_G(A_i)$ and $N_G(A_i)/A_i \cong A_7$ or $A_7 \times Z_3$ for $G \cong Mc$ or Ly, respectively.

(4) If V is a nontrivial normal elementary abelian subgroup of T then one of the following holds:

(i) $V = \langle z \rangle$.

(ii) $V = Z(J(T)) \cong E_4$ and $N_G(V) = N_G(J(T)) \leq N_G(A_i)$ or $N_G(B)$ for $G \cong Mc$ or Ly, respectively.

(iii) G is Mc, $V \leq A_i$, and $N_G(V) \leq N_G(A_i)$.

Proof. The first two remarks follow from 12.2 and the structure of Aut(Mc). Further by 12.2, $Aut_{C(z)}(A_i) \cong L_3(2)$ and $C_G(A_i) = BA_i$. So as Ly has an Mc/Z_3 subgroup containing $C_G(A_i)$, $N_G(A_i)/BA_i \cong Aut_{Mc}(A_i) \cong A_7$ if G is Ly. On the other hand if G is Mc then as G has one class of involutions and A_i is weakly closed in T by (1) and (2), $N_G(A_i)$ is transitive on $A_i^{\#}$ so as $Aut_{C(z)}(A_i) \cong L_3(2)$ it follows that (3) holds in this case. It remains to show $N(A_i)/A_i$ splits over BA_i/A_i if G is Ly. This follows as the subgroup X of $N(A_i)$ with $X/BA_i \cong A_6$ splits, and the latter remark follows from the existence of a perfect $U_4(3)/Z_3$ subgroup of Ly and lemma 10.7.4 in [4]. So (3) is established.

Assume the hypothesis of (4). As $Z(T) = \langle z \rangle$, $z \in V$. As $Z_2(T) = Z(J(T)) = E$, we may take $E \leq V$. As $C_G(E) \leq N(J(T))$, $N_G(E) = N_G(J(T))$ and by (1) and (2) the latter group normalizes A_i if G is Mc, while if G is Ly it acts on B as $B = O(C(E))$. So we may take E proper in V and then as $Z_3(T) \leq J(T)$, (1) and (2) imply $V \leq A_i$ for i = 1 or 2 and G is Mc. Now $A_i = J(C_G(V))$, so $N_G(V) = C_G(V)N_G(A_i)$. Also $C_G(V) = C_G(V) \cap C(E) \leq C_G(V) \cap N(J(T))$ $\leq N(A_i)$, so (iii) holds.

(12.7) If $Q \leq M < G$ with $F^*(M)$ simple then either G is Ly and $M \leq N_G(B)$ or G is Mc and either

(1) $M \cong M_{22}$ or $Aut(U_4(2))$. For each isomorphism type there are two such subgroups fused in Aut(G). The $Aut(U_4(2))$ subgroups are contained in a $U_4(3)$ subgroup.

(2) $M \cong U_4(3)$ or $Z_2/L_3(4)$ and M is uniquely determined by its isomorphism type.

Proof. Let $F^*(M) = L$. By 12.4 we may assume $S = T \cap M \in Syl_2(M)$.

If $L \cong A_n$ then as $|Aut(L)|_2 \geq |Q|$, $n \geq 14$ if G is M(23), $n \geq 8$ if G is Mc, $n \geq 10$ if G is Ly or M(22), and $n \geq 12$ otherwise. If $n \geq 14$, $|L|$ is divisible by 49 whereas the order of M(23) is not. If $n \geq 8$ or 10 then L contains an $A_5 \times Z_3$ or $A_5 \times A_5$ subgroup, whereas Mc, Ly, and M(22) do not contain a subgroup of the respective type. Finally if $n \geq 12$ then L contains an element of order 35, whereas the remaining groups do not. So L is not an alternating group.

Suppose L is sporadic. Evidently $|Q| \leq |S| \leq |T|$ and $|L|_2 \leq |S| \leq |Aut(L)|_2$. This restricts the possibilities for L significantly. Further the order of L divides the order of G. Finally if G is M(22) then $O_2(C_M(z))$ is of order at least 2^{10} for $z \in Z(Q)^\#$ 2-central in G. These observations leave the following possibilities:

(a) G is Mc and M is $Aut(M_{12})$, M_{22}, or J_2.

(b) G is Co_3 and M is M_{24} or Aut(HS).

(c) G is Ly and M is $Aut(M_{22})$, $Aut(J_2)$, or Aut(Mc).

(d) G is M(23) and M is Aut(M(22)).

If M is $Aut(M_{22})$, Aut(HS), or M_{24} then M possesses a dihedral subgroup of order 22 while G does not. If M is Aut(M(22)) then $S = T$ while the centers of Sylow 2-subgroups of Aut(M(22)) and M(23) are of order 2 and 8, respectively. If L is J_2 or M is $Aut(M_{12})$ then M has an $E_4 \times A_5$ subgroup while G does not. If G is Ly we are left with $M \cong Aut(Mc)$. Here its easy to see $M \leq N(B)$, whereas N(B) has no Mc-subgroup. For example $M = \langle C_M(e) : e \in E^\# \rangle$, with $E_4 \cong E \trianglelefteq T$ (by an inductive appeal to 12.7) and as $T \leq N_M(B)$, $C_M(e) \leq N(B)$, since $C_G(e) \cap N(B)$ is then the only maximal subgroup of $C_G(e)$ containing $C_M(e)$. The only remaining case is $G \cong Mc$ and $M \cong M_{22}$. The existence of such a subgroup can be seen from the embedding of Mc in Co_1 and the action of the latter group on the Leech lattice. Next $M = \langle N_M(A_i) : i=1,2 \rangle$, where $\mathfrak{A}(T) = \{A_i : i=1,2\}$,(cf. 12.6) $N_M(A_1)/A_1 \cong A_6$, and $N_M(A_2)/A_2 \cong S_5$, with an element of order 3 in $N_M(A_2)$ semiregular on A_2. Now $N_G(A_i)/A_i \cong A_7$ by 12.6, and hence contains a unique T-overgroup satisfying the conditions of $N_M(A_i)$. Further A_1 and A_2 are fused in Aut(G). So 12.7.1 holds and the treatment of the case L sporadic is complete.

So $L \in Chev(r)$ for some prime r. Suppose $r = 2$. Let $Z = Z(S)$. By 10.1, $Z \leq L$ and either Z is contained in a long root group R of L or L is $Sp_n(q)$ or $F_4(q)$, where $q = 2^m$ and $n \geq 4$.

Suppose the latter holds. The 2-share of $F_4(q)$ is at least 2^{24} and that of $Sp_n(q)$ is $2^{mn^2/4}$, so if L is either of these groups then L is symplectic and $m \leq 4$. In that event $m_p(L) \geq 2$ for $q^4 \equiv 1 \mod p$, so as neither 13^2 nor 17^2 divide the order of G, $m \leq 2$. If $mn^2/4 \geq 14$ then $|G|_2 \geq 2^{14}$ so G is a Fischer group. Then $|G|_2 =$

$|Q| = 2^{17}$ if G is M(23) while $|G|_2 = 2^{16}$ if G is M(22). We conclude (m,n) is (1,8) or (2,6) and $G \cong M(22)$ or M(23), respectively. But then the order of G is not divisible by the order of L. We are left with (m,n) = (2,4) or (1,6) and G not M(23). In the first case 17 divides the order of L but not G. In the second $|M|_2 = 2^9$ so G is not M(22). Thus $C_T(Q)$ is of order 2 so $|Z| = 2$, while as $M \cong Sp_6(2)$, $Z \cong E_4$.

So Z is contained in a long root group R of L and L is not symplectic or $F_4(q)$. Let $A = O_2(C_G(z))$. $A \le Q$ so $A \trianglelefteq C_M(z)$. Hence by 10.2 and 10.3, as L is not $A_8 \cong L_4(2)$, and as $|S| \ge 2^7$, one of the following holds:

(i) $A \le R$.

(ii) A is special.

(iii) A is elementary abelian and $L \cong L_n(q)$.

(iv) $L \cong {}^2F_4(q)$ or the Tits group.

(v) $L \cong L_3(4)$.

Now from 12.2, A is not special. In case (iv) the 2-share of M is 2^{12m} or 2^{11} so q = 2 and G is M(22). As $Q \trianglelefteq C_M(z)$ for z 2-central, the 2-local structure of the Tits group supplies a contradiction. In case (v) $|S| \le 2^8$ so G is Mc or Ly. Further $M = \langle N_G(A_i) : i=1,2\rangle\langle t\rangle$ where $\mathfrak{A}(T) = \{A_i : i=1,2\}$, t =1 if G is Mc and $A_1^t = A_2$ if G is Ly, and $N_M(A_i)^\infty/A_i \cong L_2(4)$ acts naturally on A_i. These conditions determine a unique $N_T(A_i)$-overgroup in $N_G(A_i)$ (cf. 12.6) so M is uniquely determined. Moreover if G is Ly then $N_M(A_i) \le N(B)$ so $M \le N(B)$. Hence the lemma holds in this case.

We are left with (i) or (iii), so A is elementary abelian. Hence from 12.2, G is Mc, Ly, or Co_3. Further $C_M(z)$ is a T-overgroup of $C_G(z)$ with $O_2(C_M(z))$ special. Thus $(G,C_M(z))$ is $(Ly,S_5/Q_8{*}D_8)$, $(Ly,(S_3 wr Z_2)/(Q_8{*}Q_8))$, $(Mc,E_4/(SL_2(3){*}SL_2(3)))$, or $(Mc,S_3/\text{special}(64))$, from the nature of these overgroups. This forces G = Mc and M = $Aut(U_4(2))$ or $Z_2/L_3(4)$ by the structure of the centralizer of a long root element in Aut(L). The latter case has already been handled so assume the former. Again $M = \langle N_M(A_i) : i=1,2\rangle$ with $N_M(A_j)/A_j \cong S_5$ acting naturally on A_j and $N_M(A_{3-j})/A_{3-j} \cong S_4$, for j = 1 or 2. As usual this determines $N_M(A_j)$ up to the choice of j, and as $N_M(A_{3-j}) \not\le N(A_j)$, $N_M(A_{3-j})$ is also determined. So there are just two choices for M, and these groups are fused in Aut(G). Thus 12.7 holds in this case since a $U_4(3)$ subgroup of Mc contains two T-overgroups isomorphic to $U_4(2)$.

We are reduced to the case r odd. $m_2(Q) > 3$, so L is not $L_2(q)$ or $L_3^\epsilon(r)$. Hence a Sylow r-group of L is nonabelian of order at least r^4. It follows that either r = 3 and G is not Ru or ON, or G is Ly and r = 5. By 9.6, there is an S-invariant subgroup X which is the central product of subgroups $X_i \cong SL_2(q)$, $1 \le i \le k$; without loss $z \in Z(X)$. If G is Ly and $r = 5$ there is no such T-invariant subgroup of $C_G(z)$. So $r = 3$. The existence of The S-invariant X forces $q = 3$ and $G \cong$ Mc or Ly with $C_M(z)$ the centralizer of an involution in $U_4(3)$ or $Z_2/U_4(3)$, respectively. It follows that either G is Mc and L is $U_4(3)$ or $PSp_4(3)$, or G is Ly and L is $U_4(3)$. As $PSp_4(3) \cong U_4(2)$, we can ignore this case since it was handled above when $r = 2$. In the remaining cases the usual arguments show M is determined up to conjugacy in G and hence complete the proof.

The various lemmas in this section immediately yield the description of the over groups of T contained in the following two lemmas:

(12.8) If G is Ly then $\mathcal{N}(T) = \{N_G(B), C_G(z)\}$.

(12.9) If G is Mc then

$$\mathcal{N}(T) = \{N_G(A_i), M_j, C_G(z) : i = 1,2;\ 1 \le j \le 4\},$$

where $M_1 \cong U_4(3)$, $M_2 \cong Z_2/L_3(4)$, and $M_3 \cong M_4 \cong Aut(M_{22})$. $N_G(A_1)$ and $N_G(A_2)$ are fused in Aut(G) as are M_3 and M_4.

(12.10) Let G be Co_3 and set $E = \Omega_1(Z_2(T))$. Then

(1) $Z_2(T) \cong Z_4 \times Z_2$ and $E = \Omega_1(Z(C_T(E))) \cong E_4$.

(2) $N_G(E)/O_2(N_G(E)) \cong E_4/3^{1+2}$ with $O_2(N_G(E))$ special of order 2^8 with center E.

(3) T has just 4 nontrivial normal elementary abelian subgroups $(A_i : 1 \le i \le 4)$. A_i is of order 2^i.

(4) Let $A = A_4$. Then $N_G(A)/A \cong L_4(2)$ and $N_G(A_3) \le N_G(A)$.

(5) $\mathcal{N}(T) = \{C_G(z), N_G(E), N_G(A)\}$.

Proof. Let $H = C_G(z)$, $H^* = H/\langle z\rangle$. Then H is the covering group of $Sp_6(2)$, so $Z(T^*) = Z_2(T)/\langle z\rangle \cong E_4$ and contains involutions of type a_2, c_2, and b_1. (In the language of [12]) From page 88 in [12], elements of type b_1 and c_2 lift to elements of order 4 in H while elements of type a_2 lift to involutions. Hence (1) is established modulo the assertion $\Omega_1(Z(C_T(E))) \le E$, which appears at the top of page 89 of [12].

As $2|C_G(t)|_2 < |G|_2$ for t a non-2-central involution of G, $E^\# \subseteq z^G$ and then by (1), $Aut_G(E) = GL(E)$. Let $R = O_2(C_G(E))$, $X \in Syl_3(C_H(E))$, and $X \le Y \in Syl_3(N_G(E))$. As

$O_2(C_{H^*}(E^*)) = [O_2(C_{H^*}(E^*)),X]$, $R^* = O_2(C_{H^*}(E^*))$. Let $D = \langle Z_2(T)^X\rangle$. Then $D^* \cong E_8$ and $R/D \cong E_{16}$, so either $Y \cong 3^{1+2}$ is irreducible on R/E or $Y \cong E_{27}$ and Y acts on $D = C_R(C_X(D))$. The latter is impossible as Y does not act on $\Phi(D) = \langle z\rangle$. Now (2) follows.

Next $R = D*[R,C_X(D)]$, $D \cong Q_8 \times Z_2$, and $D^Y = \{D_i : 1 \le i \le 3\}$ with $[R,C_X(D)] = D_2D_3$. So $R = D*D_2*D_3$. Its now easy to see that $C_T(E) = R\langle s\rangle$ where s inverts X, and $C_{R/E}(C_T(E)) = C_{R/E}(S) = U_0/E$ with $U_0 \cong Z_4 \times Z_4 \times Z_2$, so $\Omega_1(U_0) = U$ is the unique normal E_8-subgroup of T. Further $|C_T(U):C_R(U)| = 2$ with $C_R(U) = U_0\langle r\rangle$, where r is an involution inverting U_0, and $[R,w] = U_0$ for each w in $C_T(E) - R$. Thus (3) holds with $A_1 = \langle z\rangle$, $A_2 = E$, $A_3 = U$, and $A_4 = \langle r,U\rangle$. Indeed this even shows $A = A_4$ is the unique normal E_{16}-subgroup of $C_T(E)$. Hence $N_G(A)\cap N(E)$ is transitive on $E^\#$, so as $A = C_G(A)$ and $Aut_H(A)$ is the stabilizer of z in $GL(A)$, $N_G(A)/A \cong L_4(2)$. $C_H(A_3) \le N_H(A)$ so as $Aut_G(A) = GL(A)$, $N_G(A_3) \le N_G(A)$. Thus (4) is established. 12.5, 12.7, and the first four parts of 12.10 imply 12.10.5.

(12.11) Let G be ON and $Z = O_2(C_G(z))$. Then

(1) $Z \cong Z_4$.

(2) $\langle z^G \cap T\rangle = W \cong Z_4^3$.

(3) $N_G(W)/W \cong L_3(2)$.

(4) $\mathcal{N}(T) = \{C_G(z), N_G(W)\}$.

Proof. Part (1) follows from 12.2. Part (1) implies W is abelian; cf. 2.5 in [5]. This remark makes it easy to derive (2) and (3). If M is in $\mathcal{N}(T)$ then by 12.5 and 12.7, $F^*(M) = O_2(M)$. Thus $V = \langle z^M\rangle$ is elementary abelian. For $t \in T - W$, $\Phi([t,W]) \ne 1$, so $V \le W$. But as $C_H(X) \le N(W)$ for each hyperplane X of $\Omega_1(W)$ and $GL(\Omega_1(W)) = Aut_G(\Omega_1(W))$, $N_G(X) \le N(W)$. Thus either $M \le C(z)$ or $W \trianglelefteq M$. That is (4) holds.

(12.12) Let G be Ru and $H = C_G(z)$. Then

(1) Q is of class 3, $Q/\Phi(Q) \cong E_{64}$, $\Phi(Q) \cong E_{32}$, and $C_G(\Phi(Q)) = Q_1 \cong Q_8 \times E_{16}$.

(2) H acts indecomposably on $Q/\Phi(Q)$ with $C_{Q/\Phi(Q)}(H^\infty) = Q_1/\Phi(Q)$ and with Q/Q_1 the natural module for $E(H/Q) \cong L_2(4)$.

(3) $[Q,x] = \Phi(Q)$ for $x \in Q_1 - \Phi(Q)$ and $\Phi(Q)/\langle z\rangle$ is the natural module for $E(H/Q)$.

(4) There exists unique elementary abelian normal subgroups V_i of T of order 2^i for $1 \le i \le 5$. $\Phi(Q) = V_5$ is the unique maximal normal elementary abelian subgroup of T.

(5) Let $V = V_3$. Then $C_G(V) = P$ is a special 2-group of order 2^{11} with center V and $N_G(V)/P \cong L_3(2)$.

(6) $\mathcal{N}(T) = \{C_G(z), N_G(V)\}$.

Proof. Heavy appeals will be made to Parrott [33]. In particular (1) is a restatement of lemmas 1 and 2 of [33]. Let $X \in Syl_3(H)$. Then from page 25 of [33], $Q_8 \cong C_Q(X) \leq Q_1$, so Q/Q_1 and $\Phi(Q)/\langle z\rangle$ are the natural module for $E(H)/Q \cong L_2(4)$. Now the commutator relations listed in lemma 12 of [33] complete the proof of (2) and (3).

If U is a normal elementary abelian subgroup of T contained in $\Phi(Q)$ and $W/U = Z(T/U)$ then $W \leq C(Q/\Phi(Q)) = Q$. Then by (3) either $W \leq \Phi(Q)$ or $\Phi(Q) \leq [W,Q] \leq W$. In the latter case $W \leq \Omega_1(C_Q(\Phi(Q))) = \Phi(Q)$ by (1), so $W = \Phi(Q)$. In the former the representation of H on $\Phi(Q)$ shows W is determined by its order. Hence (4) is established. Part (5) is a restatement of lemma 11 of [33].

Notice $C_G(V_2) = C_T(V_2) \leq N_G(V)$ and $Aut_{N_G(V)}(V_2) = GL(V_2)$, so $N_G(V_2) \leq N_G(V)$. Further $N_G(V)$ is transitive on hyperplanes of V, so if $g \in G$ with $V \cap V^g$ noncyclic then $g \in N_G(V)$. Hence $N_G(V_4) \leq N_G(V)$. Finally $N_G(V_5) \leq C_G(z)$ as $\langle z\rangle = \Phi(C_G(V_5))$ by (1). Hence (6) holds given 12.5 and 12.7.

(12.13) Let G be M(22) or M(23) and D the set of 3-transpositions in G. Then

(1) If t is an involution in G which is the product of $m \leq 3$ commuting members $D(t) = \{d_i : 1 \leq i \leq 3\}$ of D then the set D(T) is uniquely determined so that $C_G(t) \leq N_G(D(t))$.

(2) If G is M(22) then $Z(T) = \langle z\rangle$ with $D(z) = \{a,b\}$ and $Z(Q) = \langle a,b\rangle$ while if G is M(23) then $Z(T) = \langle ab,d\rangle$ with a, b, and d in D.

(3) $A = \langle D \cap T\rangle \cong E_{2^m}$, with m = 10 or 11 and $N_G(A)/A \cong M_{22}$ or M_{23}, for $G \cong M(22)$ or M(23), respectively.

(4) Let $V = \langle V \cap D\rangle \trianglelefteq T$ with $D(Z(T)) = \{D(z'):z' \in Z(T)\} \subseteq V$. Then one of the following holds:

(i) $C_G(V) \leq N_G(A)$.

(ii) G is M(22) and $V = \langle a,b\rangle$ or $V = U \cong E_{32}$ is uniquely determined.

(iii) G is M(23) and $V = \langle a,b,d\rangle$ or $V = U \cong E_{64}$ is determined.

(5) Let U be the group of (4), $M = N_G(U)$, and $R = O_2(M)$. Then:

(i) R is special with $Z(R) = U$.

(ii) $M/R \cong S_3 \times A_6$ or $S_3 \times A_7$ for $G \cong M(22)$ or $M(23)$, respectively.

(iii) If G is M(23) then U is the A_7-module for M.

(iv) If G is M(22) then U is an indecomposable M-module with [U,M] the A_6-module.

(v) R/U is the tensor product of the $L_2(2)$-module with the 4-dimensional module for A_m, m = 6 or 7.

(6) If G is M(22) then $\mathcal{N}(T) = \{C_G(z), N_G(A), N_G(U)\}$.

(7) If G is M(23) then

$$\mathcal{N}(T) = \{C_G(d), C_G(ab), C_G(abd), N_G(A), N_G(U)\}.$$

Proof The first three parts are well known properties of 3-transposition groups; cf. section 13 in [13].

Assume the hypothesis of (4). If M is M(22) then as $a \in V$, $C_G(V) \le C(a)$. Let $H = C_G(a)$ and $H^* = H/\langle a\rangle$. Keep in mind $(H\cap D)^*$ is the set of transpositions in H^*. As $H^* \cong U_6(2)$ and $V \trianglelefteq C_T(a)$, either $C_G(V) \le N(A)$ or $P = C_P(V)C_T(a)$, where P^* is the unique minimal parabolic of H^* containing $C_T(a)^*$ but not contained in $N_H(A)^*$. ($N_H(A)^*$ is the $L_3(4)$-parabolic.) $P^* \le K^* = C_{H^*}(z^*)$ and $(D\cap O_2(K))^* = \langle z^*\rangle$, while a Sylow 3-group of P centralizes a unique member $t\alpha$ of $(D\cap P)\alpha$, where $K\alpha = K/O_2(K)$. Namely $\langle t\alpha\rangle = Z(P\alpha)$. So $V = \langle a,b\rangle$ or $V\alpha = \langle t\alpha\rangle$ and we may assume the latter. Now $U = \langle t^P\rangle = \langle D\cap\langle t\rangle O_2(K)\rangle \cong E_{32}$ so V = U is determined.

Similarly if G is M(23) let $H = C_G(d)$ and $H^* = H/\langle d\rangle$. Then $d \in V$ and $H^* \cong M(22)$ with $(D\cap H)^*$ the set of 3-transpositions of H^*, so (4) holds.

If G is M(22) the structure of $N_G(U)$ is perhaps best determined using the techniques of section 30. See in particular the discussion of M(22) in section 30. If G is M(23) then by induction on the order of G, $N_G(U)\cap C(d)$ is described in (5). Moreover $Aut_{N(A)}(U) = A_7$ acts naturally on U using (3) and (4). The structure of $N_G(U)$ is then easily retrieved from these observations.

Parts (6) and (7) will follow from 12.5, 12.7, and the earlier parts of the lemma once we show that if $T \le M \le G$ with $F^*(M) = O_2(M)$ then $D(Z(T)) \subseteq O_2(M)$. Assume not and let $t \in D(Z(T))-O_2(M)$, $Y = \langle D\cap M\rangle$, and $X = \langle t^Y\rangle$. Then $t \notin O_2(X)$. Now by (2) there is $s \in D(Z(T))$ with $ts \in Z(T)$. As $F^*(M) = O_2(M)$, $ts \in O_2(M)$. If $r \in (D\cap M)-t^X$ then $[r,X] = 1$ by a well known property of 3-transposition groups. Thus if $s \notin t^X$ then $[s,X] = 1$ and so as $ts \in O_2(M)$, it follows that $t \in O_2(X)$, a contradiction. So $s \in t^X$ and thus as T acts on $\{s,t\}$, $T \le N(X)$. So without loss M = XT. Similarly if G is M(23) then

$[d,X] = 1$, so $M \leq C(d)$ and we have a contradiction by induction on the order of G as $C(D)/\langle d\rangle \cong M(22)$. So G is M(22). Now $z = ts \notin Z(X)$ as $\langle t,s\rangle \trianglelefteq C(z)$. But $\langle z\rangle = Z(T)$ and $O(M) = 1$, so $C_M(X) = 1$. Hence $D\cap M = t^X$. So $|t^X \cap T| = |T\cap D| = 22$.

Let $V = \langle z^X\rangle$. As $z \notin Z(M)$, $[V,t] \neq 0$, so as $[T,t] = \langle z\rangle$, t induces a transvection on V with center $\langle z\rangle$. Also $\langle t,V\rangle \trianglelefteq T$, $X = \langle t^X\rangle$, and X is irreducible on V since $\langle z\rangle = Z(T)$ and $V = \langle z^X\rangle$. These facts together with 6.6 and the fact that $(D\cap V)V/V$ is a set of 3-transpositions of M/V force $X/V = Sp(V)$ or S_r, $r \equiv 2$ or 3 mod4. But then $|t^X \cap T| = 2k$ where k is the number of transvections in a Sylow 2-group of Sp(V) or S_r. Thus $k = 2^m-1$, where $2m = m(V)$, or $(r-\epsilon)/2$, respectively. As $11 \neq 2^m-1$ for any m, we have a contradiction in the first case. Similarly in the second case $r = 22$ or 23 and $C_M(t)$ has an A_{r-2}-section, which is not the case. Hence the proof is complete.

13. 3-structure in J_3.

In this section G is J_3 and $Q \in Syl_3(G)$. It is well known that:

(13.1) (1) $\Omega_1(Q) = A \cong E_{27}$.

(2) $Z(Q) = Z \cong E_9$ and $Q/Z \cong 3^{1+2}$.

(3) Z is a TI-set in G with $Q \trianglelefteq N_G(Z)$ and $N_G(Z)/Q \cong Z_8$ is regular on $Z^{\#}$.

(4) $N_G(X) \cong PGL_2(9)/Z_3$ for each $X \in \Gamma$, where Γ is the set of subgroups of order 3 of A not in Z.

(13.2) $\mathcal{N}(A) = \{N_G(Z), N_G(X) : X \in \Gamma\}$.

Proof. Let $M \in \mathcal{N}(A)$. By 2.2 and 13.1, $\mathcal{M}_G(A,3') = 1$. Hence $O_{3'}(M) = 1$. From 13.1, Z is strongly closed in Q and Z is a TI-set in G, so if $1 \neq Z \cap O_3(M)$ then $M = N_G(Z)$. Thus we may assume $Z \cap O_3(M) = 1$, so if $O_3(M) \neq 1$ then $M = N_G(X)$ for some $X \in \Gamma$. Thus we may take $O_3(M) = 1$. Therefore there is a component L of M and 3 divides the order of L.

As A is abelian, $A \leq N(A \cap L) \leq N(L)$. As Z is a TI-set and $Q \trianglelefteq N(Z)$, Z is faithful on L. Suppose $C_A(L) = X \neq 1$. Then $X \in \Gamma$ and $L \leq E(C_G(X)) \cong A_6$. Then as Z is faithful on L, $L = E(C_G(X)$ and by a Frattini argument, $M = LN_M(Z) \leq N_G(X)$ as $A \trianglelefteq N_G(Z)$. Thus we are done in this case.

So A is faithful on L and hence $m_3(Aut(L)) \geq 3$. As $\Omega_1(Q)$ is abelian of rank 3 and $m_p(L) \leq 1$ for each prime $p > 3$, L is $L_2(27)$. But now as Z is strongly closed in A, we have a contradiction.

14. 3-structure in F_3.

In this section G is F_3. We make heavy use of Parrott [32], to the extent that the reader should probably have [32] available while going through section 14. From section 4 of Parrott we obtain:

(14.1) (1) G has three classes of elements of order 3 with representatives d, c_1 and c_2.

(2) d is 3-central. Let $N = N_G(\langle d \rangle)$ and $Q = O_3(N)$. Then $|Q| = 3^9$ and $N/Q \cong GL_2(3)$.

(3) $N_G(\langle c_1 \rangle) \cong Z_2/(Z_3 \times (SL_2(9)/E_{81}))$.

(4) $N_G(\langle c_2 \rangle) \cong Z_2/(Z_3 \times G_2(3))$.

(5) Let $T \in Syl_3(N)$, $D = Z_2(T)$, $X = N_G(D)$, and $M = O_3(X)$. Then $\langle d \rangle = Z(T)$, $D \cong E_9$, and $|M| = 3^9$, and $X/M \cong GL_2(3)$.

The next two results are established together.

(14.2) (1) $K = Z_2(M) = [M,M] \cong E_{3^5}$.

(2) $D = Z(M)$ and $c_1, c_2 \in K$.

(3) K/D is the natural module for $X/C_X(K/D) \cong O_3(3)$ and D is the natural module for $X/M \cong GL_2(3)$.

(4) M/D is special.

(5) X acts indecomposibly on $M/K \cong E_{81}$ and $T/K \cong Z_3 \times (Z_3 wr Z_3)$.

(6) $\langle d \rangle^G \cap K = \langle d \rangle^X \cup \langle c_3 \rangle^X$, with the orbits of length 4 and 36.

(14.3) (1) $D_3 = Z_2(Q) \cong E_{27}$ and $D_3/\langle d \rangle$ is the natural module for N/Q.

(2) $Q' = [Q,Q] = Z_3(Q)$, $D_3 = Z(Q')$, and $D_3/\langle d \rangle$ is the only noncentral $O^{3'}(N)$-chief factor in Q'. $Q' \cong E_9 \times S$, where S is the nonabelian group of order 27 and exponent 9.

(3) N is indecomposable on $Q/Q' \cong E_{81}$ and $T/Q' \cong Z_3 \times (Z_3 wr Z_3)$.

(4) $D_{23} = \langle D, c_2, c_3 \rangle \trianglelefteq N$. $D_{23} \cong E_{81}$ and $C_Q(c_2) = C_Q(D_{23}) \cong E_{27} \times 3^{1+2}$ with $\langle d \rangle = \Phi(C_Q(D_{23}))$. $C_3 = C_G(D_3) = C_Q(D_{23})Q'$.

(5) $Q/D_{23} \cong E_9 \times 3^{1+2}$.

(6) $C_Q(D_{23}) = \Omega_1(C_3)$.

Now for the proof of 14.2 and 14.3. First the proof of 14.2. By Parrott [32], page 397, (i) and (ii), there is $K \trianglelefteq X$ with $E_{3^5} \cong K$ and $M/K \cong E_{81}$. Moreover from 4.1 in [32], the actions of X on D and K/D are described in 14.2.3, and $K = \langle c_1,c_2,c_3,D\rangle$, with c_iD representatives for the orbits of X on K/D and with $c_3 \in d^G$. Define $C_i = C_M(c_i)$, $1 \leq i \leq 3$. From [32], 4.3.(ii), $|M:C_i| = 9$ and $\langle c_i,D\rangle = Z(C_i)$, $1 \leq i \leq 3$. Then $|C_i| = |C_G(c_i)|_3$ for i = 1 and 2, so $C_i \in Syl_3(C_G(c_i))$. We see also that $D = Z(M)$.

As X is irreducible on K/D, $K \leq Z_2(M)$. As $|M:C_i| = |D|$ and as c_iD, $1 \leq i \leq 3$, are representatives for the orbits of X on $(K/D)^{\#}$, we conclude $[k,M] = D$ for each $k \in K - D$. Then $D \leq [M,M]$ and as M/K is abelian and N is irreducible on K/D, we have $[M,M] = K$ or D. By 14.1.4 and the structure of a Sylow 3-group of $G_2(3)$, C_2/D is nonabelian. Hence $[M,M] = K$.

As c_iD, $1 \leq i \leq 3$, are representatives for the orbits of X on K/D with $c_3 \in d^G$ and $[c_i,M] = D$, we conclude from 14.2.3 that 14.2.6 holds.

Next by 4.3.(ii) in [32] an involution in $Z^*(X)$ is fixed point free on M/K, so X has two chief factors on M/K, each of rank 2. Hence either X is indecomposable on M/K, in which case $T/K \cong Z_3 \times (Z_3 wr Z_3)$, or M/K is the sum of two irreducibles and $t \in T - M$ is quadratic on M/K. But from a remark in the last paragraph of the proof of 4.3 of [32] (which is proved below), an element c_4 in T - M is not quadratic on M/K, so 14.2.5 holds. To complete the proof of 14.2 it remains to show $K/D = Z(M/D)$. But if not from the action of X on M/K, $|M:Z| = 9$ where $Z/D = Z(M/D)$. Then $M = \langle Z,m_1,m_2\rangle$ so $[M/D,M/D] = \langle [Dm_1,Dm_2]\rangle \cong Z_3$, whereas we have seen already that $[M,M] = K$.

We next establish 14.3. We find at the bottom of page 398 of [32] that $D_{23} = \langle D,c_2,c_3\rangle$ and C_3 are normal in N. Hence $D_3 = \langle c_3,D\rangle = Z(C_3) \trianglelefteq N$. Next from 14.1.4, $C_N(c_2) = \langle c_2\rangle \times N_2$ with N_2 a maximal parabolic of $G_2(3)$. In particular $C_Q(c_2) \cong E_{27} \times 3^{1+2}$ with $\langle d\rangle = \Phi(C_Q(c_2))$ and N_2 acts indecomposably on $C_Q(c_2)/\langle d,c_2\rangle$. Thus $D_3 \leq Z(C_Q(c_2))$ as D_3 is a normal E_{27}-subgroup of $C_N(c_2)$. So $C_Q(c_2) = C_Q(D_{23}) \trianglelefteq N$ and $D_3/\langle d\rangle$ is the natural module for N/Q.

Near the bottom of page 398 of [32] we find there is an involution u in $Z^*(N)$ inverting elements c_4 and c_{12} of N defined in [32], so both are in Q. Then as $|Q:C_3| = 9 = |\langle c_4,c_{12},C_3\rangle|$, we have $Q = \langle c_4,c_{12},C_3\rangle$. Also u inverts Q/C_3, so that section is the natural module for N/Q. (Incidentally as $c_{11} \in T - Q$, this shows $[c_{11},c_4] \in$

$(c_{12})^{\epsilon}c_3$, ϵ = +1 or −1, while by 4.3.(iii) of [32], $C_3/K = C_{M/K}(c_4)$, so we have our proof that c_4 is not quadratic on M/K.)

Let $Q_0 = \langle c_{31}, D_{23}\rangle$. At the bottom of page 398 of [32] we find $Q_0/D_{23} = \Phi(Q/D_{23})$. Thus as $c_{31} \in C_3$, $C_3 = C_Q(c_2)Q_0$ by an order argument. Also $Q/D_{23} \cong E_9 \times 3^{1+2}$ with $C_3/D_{23} = Z(Q/D_{23})$, and N has two chief factors C_3/Q_0 and Q/C_3 on Q/Q_0, both of rank 2. So again either N is indecomposable on Q/Q_0 or Q/Q_0 is the sum of two natural modules and hence T/Q_0 is of exponent 3. But in the penultimate line of the proof of 4.3 in [32] we find there is $t \in T$ with $t^3 \in c_1Q_0$, so T/Q_0 is not of exponent 3. Hence 14.3.3 is established, once we show $Q_0 = Q'$.

As Q/Q_0 is abelian, $Q' \leq Q_0$, while as Q/D_{23} is not, $Q' \not\leq D_{23}$. As $|Q:C_Q(c_2)| = |D_3|$, $[Q,c_2] = D_3$. So $D_3 \leq Q'$. Thus either $Q_0 = Q'$ or $c_2 \notin Q'$. In that event $Q/Q' = [Q/Q',u] \oplus C_{Q/Q'}(u)$ with $C_{Q/Q'}(u) = \langle c_2, Q'\rangle/Q'$. But then $c_2 \notin [T,T]$, whereas $c_2 \in K = [M,M]$. So $Q' = Q_0$.

As N is irreducible on $D_3/\langle d\rangle$, $D_3 \leq Z_2(Q)$. We saw in paragraph 2 of this proof that $D = [c_1,M]$. Further by [32], $M = C_1C_3$. So also $D = [C_3,c_1]$. By a previous paragraph $D_3 = [Q,c_2]$. So $D_3 = [Q,c_2]$, so $D_3 = K \cap Z_2(Q)$. Then as N is irreducible on $C_Q(c_2)/D_{23}$, $C_Q(c_2) \cap Z_2(Q) = D_3$. Next $[c_{31},c_{32}] \in D_3 - D$ by a remark near the top of page 399 of [32]. We have seen that $\Phi(C_Q(c_2)) = \langle d\rangle$ and $C_3 = \langle c_{31}, C_Q(c_2)\rangle$. By [32], $c_{32} \in C_Q(c_2)$. We conclude $D_3 = C_3 \cap Z_2(Q)$. Finally as $C_3/C_Q(c_2) = \Phi(Q/C_Q(c_2))$, $Z_2(Q) \leq C_3$, so $D_3 = Z_2(Q)$.

As u centralizes Q'/D_3 and inverts Q/Q', $Q = [Q,u]$ centralizes Q'/D_3. Therefore $Q' \leq Z_3(Q)$. As K/D is the $O_3(3)$-module for $X/C_X(K/D)$, we have $[c_4,c_1] \in D_{23} - D_3$. Thus $c_1 \notin Z_3(Q)$, and hence as N is transitive on $(C_Q(c_2)/D_{23})^{\#}$, we have $D_{23} = Z_3(Q) \cap C_Q(c_2)$. Therefore $Q' = C_3 \cap Z_3(Q)$, and then as $C_3 \trianglelefteq N$, 14.3.3 implies $Q' = Z_3(Q)$.

It remains to show $\Omega_1(C_3) = C_Q(c_2)$. As N is transitive on $(C_Q(c_2)/D_{23})^{\#}$ and acts on Q'/D_{23} with $C_3 = Q'C_Q(c_2)$ and $Q' = \langle c_{31}, D_{23}\rangle$, we must show the members of $c_{31}D_{23}$ and $c_1c_{31}D_{23}$ are not of order 3. But at the top of page 399 of [32], Parrott shows T is transitive on $c_{31}D_{23}$ and on $c_1c_{31}D_{23}$. Further in the second and third paragraphs on page 399 of [32], Parrott shows c_{31} is conjugate in G to f_1 and $c_{31}c_1$ is conjugate to f_2. Finally by definition of f_i at the beginning of section 2 of [32], f_i is of order at least 9.

Thus the proof of 14.2 and 14.3 is at last complete.

(14.4) (1) $X = N_G(K)$.

(2) $N = N_G(D_{23}) = N_G(D_3)$.

Proof. First $D_3 \le D_{23}$ so $C_Q(D_{23}) = C_G(D_{23})$ by 14.3.4. Then by 14.3.4, $\langle d\rangle = \Phi(C_Q(D_{23}))$ so $N = N_G(D_{23})$. Next by 14.3.4, $C_3 = C_G(D_3)$ while by 14.3.6, $C_Q(D_{23}) = \Omega_1(C_3)$, and finally by 14.3.4, $D_{23} = Z(C_Q(D_{23}))$. Therefore $N_G(D_3) \le N_G(D_{23}) = N$ and then $N_G(D_3) = N$ as $D_3 \trianglelefteq N$.

Assume $Y = N_G(K) \ne X$. Let $\Delta = \langle d\rangle^G \cap K$. Then by 14.2.6, X has two orbits α and β on Δ, where $\alpha = \langle d\rangle^X$ is of order 4 and β is of order 36. Thus as $\langle\alpha\rangle = D$, Y is transitive on Δ and D is a TI-set under the action of Y. Also $\Gamma = \alpha^Y$ is a system of imprimitivity for Y on Δ with $|\Gamma| = 10$. For $y \in Y - X$, $N_M(D^y) = C_M(E)$ for some subgroup E of D^y of order 3. So as $|M:C_M(E)| = 9$, M is transitive on $\Gamma - \{\alpha\}$. Therefore Y is 2-transitive on Γ, so as $E_9 \cong M^\Gamma \trianglelefteq X^\Gamma$, we conclude $O^{3'}(Y^\Gamma) = L_2(9)$. This is impossible as X^Γ involves $SL_2(3)$.

(14.5) If V is a Q-invariant elementary abelian 3-subgroup of G then $V = 1$, $\langle d\rangle$, D_3, or D_{23}, or $V \in D^N$ or K^N.

Proof. Assume V is a counterexample. As $\langle d\rangle = C_G(Q)$, $d \in V$. Let $U = C_V(D_{23})$. As K^N is the set of maximal abelian subgroups of $C_G(D_{23})$ we may take $U \le K$. By 14.2.3, $[K/D,Q] = D_{23}/D$ and $C_{K/D}(Q) = D_3/D$ so:

(i) Either $U \le D_3$ or $D_{23} \le UD$.

(ii) Either $U \le D$ or $D_3 \le UD$.

Indeed $M = (Q\cap M)C_2$ as $|M:Q\cap M| = 3 = |C_2:C_2\cap Q|$, so $[c_2,Q\cap M] = [M,c_2] = D$. Now if $D_{23} \le UD$ there is $u \in (c_2D)\cap U$, so that $D = [u,Q\cap M] \le V$. That is:

(iii) Either $U \le D_3$ or $D_{23} \le U$.

Next if $D_{23} \le U$ then $V = C_V(D_{23}) = U$. Then as $|K:D_{23}| = 3$, $V = D_{23}$ or K, contrary to the choice of V as a counterexample. Therefore:

(iv) $U \le D_3$.

Similarly if $V \le D_3$ then either $V = \langle d\rangle$ or D_3, or $V \in D^N$, contrary to the choice of V. Hence we conclude from (iv) that:

(v) $V \not\le C_G(D_{23})$.

If $V \not\le Q$ then $[V,Q/C_3] \ne 1$, so $V\cap Q \not\le C_3$. But $Q/C_Q(D_{23}) \cong 3^{1+2}$ with $C_3/C_Q(D_{23}) = \Phi(Q/C_Q(D_{23}))$. Thus if $V\cap Q \not\le C_3$ then also $V\cap C_3 \not\le C_Q(D_{23})$. So if $V\cap C_3 \le C_Q(D_3)$ then $V \le C_3$ and hence (v) and 14.3.6 supply a contradiction. Thus we have shown:

(vi) $V\cap C_3 \not\leq C_Q(D_{23})$.

But now as $\Phi(V) = 1$, (vi) and 14.3.6 supply a contradiction, completing the proof of 14.5.

(14.6) $\mathcal{N}(Q) = \{N\} \cup X^N$.

Proof. Assume $Y \in \mathcal{N}(Q)$ with $N \neq Y \notin X^N$. By 14.4 and 14.5, $O_3(Y) = 1$. As $D \leq Q$, as X is transitive on $D^{\#}$, and as $\langle d\rangle = C_G(Q)$, 2.2 says $\mathcal{U}_G(Q,3') = 1$. Thus (G,Q) satisfies the hypothesis of section 8 by 2.1. Therefore by 8.1, $F^*(Y) = L$ is a nonabelian simple group.

Without loss $S = T\cap Y \in Syl_2(Y)$. Then $Z(S) = \langle d\rangle \cong Z_3$ and $O_3(C_Y(d))$ has class at least class(Q) = 4. It follows that L is not a sporadic group. Similarly if $L \cong A_n$ then as $\langle d\rangle = Z(S)$, 7.3 implies $n = 3^m+k$, $0 \leq k \leq 2$. As $O_3(C_Y(d))$ has class at least 4, $m \geq 3$. But then $m_5(Y) > 2 = m_5(G)$.

So $L \in Chev(r)$ for some prime r. If $r = 3$ then as $\langle d\rangle = Z(S)$, d is a long root element of L. As $O_3(C_Y(d))$ is of class at least 4, this is impossible. So $r \neq 3$.

Now from 9.1 and the structure of N, there is a Q-invariant abelian normal subgroup Q_0 of $N_0 = N\cap L_0$, where L_0 is the subgroup of Y inducing inner-diagonal automorphisms on L, with N_0/Q_0 is isomorphic to a subgroup of the center j(L) of the univeral group L. As j(L) and $Aut(L)/L_0$ have cyclic Sylow 3-subgroups, Q/Q_0 is of sectional rank at most 2. So as $m(Q) \geq 5$, $m(Q_0) \geq 3$.

Let $V = \Omega_1(Q_0)$. Then V is described in 14.5 and as $m(V) \geq 3$, we may take $V = K$, D_{23}, or D_3. If $V = K$ then $Q_0 \leq C_G(K) = K$, contradicting $m(Q/K) = 3$. Similarly if $V = D_{23}$ then K^N is the set of maximal abelian subgroups of $C_G(V)$, so we have the same contradiction.

Therefore $V = D_3$. Let $C_Q(D_{23}) = Q_2$. By 14.3, $Q_2 = \Omega_1(C_G(D_3))$ is of exponent 3. Thus $V = Q_0\cap Q_2$, so that Q_2/V has sectional 2-rank at most 2. As $Q_2/D_3 \cong E_{27}$ we have a contradiction.

15. Some subgroups of the Monster

In this section $F = F_1$ is the Monster. Let z and t be 2-central and non-2-central involutions in F, respectively, let $H = C_F(t)$, and let $H^* = H/\langle t\rangle$. Then $H^* \cong F_2$ and there is an involution s in F such that $C_{H^*}(s^*) \cong Z_2/{}^2E_6(2)/Z_2$. $C_F(z) \cong Co_1/Q^{12}$ and hence contains no ${}^2E_6(2)$ section, so all involutions in $E = \langle t,s\rangle$ are in t^F and indeed $N_G(E)$ is transitive on $E^{\#}$. It follows that $C_F(E)$ is quasisimple and $Aut_F(E) \cong S_3$. From the structure of $Aut({}^2E_6(2))$ there is an element x of order 3 faithful on E with $C_F(E\langle x\rangle) \cong \Omega^-_{10}(2)$ and an involution r in F inverting x and inducing a transvection on $C_F(E)\langle x\rangle$.

Next let $X_1 \cong A_5$ be the subgroup of $C_F(E)$ covering a subgroup of ${}^2E_6(2)$ generated by a short root group and its negative. I'll refer to F-conjugates of X_1 and H^*-conjugates of X_1^* as *root*-A_5's. From the Chevalley commutator relations, $C_F(EX_1) = E \times X_1 \times X_2$, with $X_2 \cong L_4(2)$ and with x chosen to centralize X_1X_2.

Let x_5 of order 5 in X_1. As X_2 centralises $\langle x_5,s\rangle$ it follows from the centralizers listed in [23] and an order argument that $N_{H^*}(\langle x_5^*\rangle) \cong (Z_4/Z_5) \times Aut(HS)$ and s induces an outer automorphism on $E(C_H(x_5)$ with $C_{H^*}(\langle x_5^*,s^*\rangle) \cong S_8$. It follows that:

(15.1) Let $G = F_2$. Then G contains a subgroup $M \cong Z_2 \times \Omega^-_{10}(2)$ such that the normalizer in G of some root A_5 subgroup X_1 of G is contained in M. If V is the orthogonal space for the $\Omega^-_{10}(2)$ factor then X_1 stabilizes a nondegenerate subspace of V of dimension 4 and sign -1. $N_F(X_1) \cong Z_2 \times S_5 \times S_8$.

As the order of $\Omega^-_{10}(2)$ divides that of $C_F(\langle x,r\rangle)$ it follows from the centralizers listed in [23] that $N_F(\langle x\rangle) \cong M(24)/Z_3$ and $C_F(\langle x,r\rangle) \cong M(23)$. Thus we have shown:

(15.2) M(24) contains a D-subgroup isomorhic to $\Omega^-_{10}(2)$ and M(23) contains a D-subgroup isomorphic to $Sp_8(2)$.

Given sets $\mathfrak{A}_i$, $i = 1,\ldots,n$, of groups and a group G let $\Gamma(\mathfrak{A}_1,\ldots,\mathfrak{A}_n,G)$ consist of those subgroups $A_1 \times\ldots\times A_n$ of G with A_i in $\mathfrak{A}_i$ for each i. Moreover in the remainder of this section let $\mathfrak{A}$ be the collection of all nonabelian simple groups of order divisible by 5.

(15.3) Let $G = F_2$, U a subgroup of G of order 7, $\mathfrak{B}$ the collection of quasisimple groups of order divisible by 7, and $\Gamma = \Gamma(\mathfrak{B},\mathfrak{B},G)$. Then

(1) $N_G(U) \cong Z_2/((Z_3/Z_7) \times (Z_2/L_3(4)/Z_2))$ with $E(N_G(U)) \cong L_3(4)/Z_2$ and with the outer automorphism of $L_3(4)$ which centralizes U acting as a field automorphism.

(2) Up to conjugacy in G, Γ contains a unique maximal member $L \times K$.

(3) $L = E(N_G(U)$ and $K \cong L_3(2)$. $L = E(N_G(K))$ and $K = E(N_G(L))$.

(4) If $LK \neq A \times B \in LK\cap\Gamma$ then $B = K$ and $A \cong L_3(2)$.

(5) If e is an involution in G inverting U and $A_5 \cong X \leq L$ is e-invariant then X is a root A_5 subgroup of G, $N_G(X) \leq C_G(Z(L))$, and $N_G(X)\cap C_G(U)$ is maximal in $C_G(U)$.

(6) There are two $N_G(L)$-classes of $L_3(2)$-subgroups of L. Indeed if $L_3(2) \cong A \leq L$ then either $N_G(AK) = \langle f\rangle(K \times A) \times Z(L)$ with f inducing an outer automorphism on K and A, or A is conjugate to K in $N_G(AK)$ and $N_G(AK) \cong E_4/(L_3(2) \times L_3(2) \times E_4)$ with $Z(N_G(AK))$ generated by an involution i such that $C_G(i) \cong Z_2/(F_4(2) \times E_4)$.

Proof. The first remark follows from the list of normalizers in [23]. Let $A \times B \in \Gamma$; without loss $U \leq B$. Hence $A \leq L = E(N_G(U))$. Let $\langle s\rangle = Z(L)$; by a now familar order argument s is a non-2-central involution of G. Let $V \in Syl_7(A)$. Then s centralizes A and acts on $D = O^{7'}(C_G(A))$. $B \leq D \leq E(C_G(V))$ so $B \cong L_3(2)$ or $B = D = E(C_G(V))$ and in the former case either $B = D$ or $D = E(C_G(V))$. Therefore as s centralizes U and s acts on D, either s centralizes D or $D = E(C_G(V))$ and s induces a field automorphism on D. In this last case certainly $s \notin C(D) \geq A$ so by symmetry between (A,B) and (D,A) we conclude $A \cong L_3(2)$, and replacing (A,B) by (B,A) we are reduced to the case $B \cong L_3(2)$. So without loss $AB \leq C_G(s) \cong Z_2/{}^2E_6(2)/Z_2$. Let $M = E(C_G(s))$ and $M^* = M/\langle s\rangle$. Then M^* has a subgroup $X^* \times Y^* \cong L_3(2) \times L_3(4)$ with X^* and Y^* generated by long and short root groups, respectively, and with $X^* = E(C_G(Y^*))$ and $Y^* = E(C_G(X^*))$. Hence $Y = E(C_G(U_1)$ for $U_1 \in Syl_7(X)$, so conjugating in $C(s)$ if necessary, $L = Y$ and $X = E(C_G(L))\langle s\rangle$. As long root elements of M^* lift to involutions, $X \cong L_3(2) \times Z_2$. So $K = E(N_G(L)) = E(X) \cong L_3(2)$.

Next $N_G(L)$ has two orbits on its $L_3(2)$-subgroups with representives A_1 and A_2 where $V \leq A_i$ and A_1 is centralized by an element inducing a field automorphism on L. As U is fused to V in $N_G(UV)$, $K = C_G(E(C_G(U))$ is fused to $A_0 = C_G(E(C_G(V)) \leq L$.

Notice $A_0 = C_L(O_2(C_G(UV))) = A_1$. From these remarks we also see that $N_G(KA_2) \leq N(K)$, as otherwise A_1 is conjugate to A_2 in $N_G(UV)$ and hence in $N_G(U)$, since $U = C_{UV}(A_i)$. As involutions in K^* and L^* are long and short root elements of M^*, respectively, K is not conjugate to A_0 in C(s), which shows the involution i of 15.3.6 is of the form $i = s(si)$ with $si \in s^G$. This together with the tables in [23] determines the isomorphism type of $C_G(i)$ and completes the proof of (2)-(4) and (6).

Choose notation as in (5) and observe that by (1) e acts nontrivially on the Dynkin diagram of L. However L has exactly 3 classes of A_5-subgroups and the only class invariant under an automorphism nontrivial on the Dynkin diagram is the one generated by root subgroups of L. Part (5) now follows from this observation; by 15.1, $C_G(Z(L)) = O_2(N_G(X))$.

(15.4) Let $X_{11} \leq F$ of order 11, let $\mathcal{B}$ be the collection of quasisimple groups of order divisible by 11, and let $\Gamma = \Gamma(\mathcal{B},\mathcal{B},F)$. Then

(1) $N_F(X_{11}) \cong Z_2/((Z_5/Z_{11}) \times M_{12})$.

(2) Up to conjugation, Γ has a unique maximal member $L \times K$.

(3) $L = E(N_F(X_{11}))$ and $K \cong L_2(11)$.

(4) $K = E(N_F(L))$, $L = E(N_F(K))$, and $N_F(L) = \langle e \rangle LK$ where e induces an outer automorphism on L and K.

(5) If $LK \neq A \times B \in LK \cap \Gamma$ then $B = K$ and $A \cong M_{11}$ or $L_2(11)$.

Proof. As usual the first remark can be found in [23]. Let $X = X_{11}$ and $L = E(C_F(X))$; let Y of order 5 in $C(L) \cap N(X)$ and $M = N_F(Y)$. Then $M \cong Z_2/(D_{10} \times F_5)$. In particular there is an involution u inverting Y and centralizing $E(M) \geq L$. Hence $K = \langle XY,u \rangle \leq C(L)$. Let W be of order 11 in L. Then $K \leq C(W) \cong Z_{11} \times M_{12}$ so from the list of maximal subgroups of M_{12} in [19] it follows that $K \cong L_2(11)$, M_{11}, or M_{12}, and indeed $I = C_F(L)$ is also one of these groups.

Now let $A \times B \in \Gamma$; without loss $X \leq B$ and $A \leq L$. As A and B are subgroups of M_{12} they are among $L_2(11)$, M_{11}, and M_{12}. In particular $X \leq V \leq B$ with V of order 55 and as V centralizes $A \leq L$ it follows that $V = XY$. Also if $u \in B$ and B is not $L_2(11)$, then there is $v \in N_B(Y)$ with $v^2 = u$. But then from the structure of M, v induces an involutory outer automorphism on $E(M) \cong F_5$, impossible as such an automorphism centralizes no element of order 11. As $u \in K \leq I$, this shows $K = I \cong L_2(11)$. Also as $B' = \langle u,K \rangle \leq C(A)$ it follows that $B = B' = K$. With these observations it is easy to complete the proof of the lemma.

(15.5) Let $G = F_5$. Then

(1) G has exactly two classes L_i^G, $i = 1,2$, of subgroups $L_i \cong M_{12}$. L_1 is fused to L_2 in Aut(G).

(2) $N_G(L_i) \cong Aut(M_{12})$.

(3) L_i is contained in an S_{12} subgroup of Aut(G).

Proof. The proof of the last lemma shows (in the notation of that proof) that $F_5 \cong E(M^*)$ contains $N_M(X) \cong Aut(M_{12})$.

So let $M_{12} \cong L \leq G$. There is U of order 3 in L with $O^{3'}(N_L(U)) = W \cong Z_3 \times A_4$. This forces U to be non-3-central in G as the 2-rank of the centralizer of a 3-central element is 1. Hence $Z_2/(Z_3 \times A_9) \cong N_G(U) \leq A \leq G$ with $A \cong A_{12}$ and $N_{Aut(G)}(A) \cong S_{12}$. (cf. Prop. 2.13 in [39].) Of course U moves exactly 3 of the set Δ of 12 points permuted by A. Similarly since $U \not\leq [P,P]$, where $P \in Syl_3(N_G(U))$, we conclude that up to conjugacy there is a unique subgroup V of order 3 in W which is 3-central in G and that V moves exactly the 9 points of Δ fixed by U. Its now easy to see W is determined up to conjugacy in $N_G(U)$, that the set $\theta = \theta(U,UV)$ of N(U)-conjugates of W containing V is of order 9, and that the subgroup N of G fixing W and each subgroup of UV of order 3 is isomorphic to Z_2/E_{27} and has orbits of length 1,2,3,3 on θ. Notice that if $\mathcal{W}$ denotes the set of $(Z_3 \times A_4)$-subgroups of M_{12}-subgroups of G, then we have shown that $W^G = \mathcal{W}$.

Next, up to quasiequivalence, M_{12} has a unique permutation representation of degree 12, while $Aut(M_{12})$ has no such representation. So A contains two classes L_i^A, $i = 1,2$, of M_{12}-subgroups, and these classes are fused in $N_{Aut(G)}(A)$.

Next from [39] there is an involutory automorphism α of G centralizing W with $\langle A,\alpha\rangle \cong S_{12}$. From this fact we calculate $N\langle\alpha\rangle$ has orbits of length 1,2,6 on θ. Now for $h \in N_L(UV) - N(U)$, $L = \langle W,W^h\rangle$, so the class of L under $N\langle\alpha\rangle$ is determined by the orbit of W^h under $N\langle\alpha\rangle$ in θ^h. Claim there is a unique such orbit with $\langle W,W^h\rangle \cong M_{12}$. This will show Aut(G) is transitive on its M_{12}-subgroups. Further recall there is an $Aut(M_{12})$-subgroup B of G. If B were determined up to conjugacy in G then by a Frattini argument B centralizes an involutory outer automorphism of G, which is not the case. So there are two G-classes of M_{12}-subgroups. This also shows $\mathrm{Aut}(M_{12}) \cong N_G(L_i)$ and hence $L_1 \notin L_2^G$ as L_1 is fused to L_2 by an element of $\mathrm{Aut}(G) - G$ normalizing A. Thus the proof of the lemma is complete modulo the claim.

It remains to prove the claim. Let $K = \langle W, W^h \rangle$. If W^h is in an orbit of length 1 or 2, then $O_3(N)$ acts on W^h, and hence also on K. So as $m_3(\mathrm{Aut}(M_{12})) = 2$, $C_{O_3(N)}(K) \neq 1$. However by [23], no element of order 3 centralizes an M_{12}-subgroup. Thus $K \cong M_{12}$ precisely when the orbit of W^h is of length 6, completing the proof.

(15.6) Let $G \cong F_2$. Then

(1) Up to conjugacy G contains a unique subgroup $L \cong A_{12}$.

(2) $N_G(L) \cong Z_2 \times S_{12}$.

(3) Let $A_5 \cong X \leq L$ move exactly 5 of the 12 points permuted by L. Then X is a root A_5-subgroup of G.

Proof. The natural module for S_{12} supplies an embedding of S_{12} in $O_{10}^-(2)$. (cf. Exercise 7.7 in [6]) So by 15.1, G contains an $S_{12} \times Z_2$ subgroup.

Conversely let L be an A_{12}-subgroup of G and $X \cong A_5$ a subgroup of L moving 5 of the 12 points permuted by L. There is a subgroup of order 7 of $C_L(X)$ inverted by an involution of L acting on X, so by 15.3.5, X is a root A_5-subgroup of G. Let $A_6 \cong Y \leq C_L(X)$. Observe $E(C_L(Y)) = E(C_G(Y))$; as $E(C_L(Y)) \cong A_6$ this follows from 15.1 and the fact that Y contains an L-conjugate of X. Hence $L = \langle B, E(C_G(Y)) \rangle$, where $B = E(C_L(X)) \cong A_7$. But by 15.1 $C_G(X)^\infty = D \cong A_8$. Thus B is determined up to conjugacy in D and Y is determined up to conjugacy in B, so L is determined up to conjugacy in G. This completes the proof of (1) and (3) while (2) is an easy consequence of 15.1.

(15.7) Up to conjugacy F contains a unique subgroup $X \times M$ with $X \cong A_5$ and $M \cong A_{12}$. X is a root A_5-subgroup of F and $N_F(X) = N_F(M) = XM\langle e \rangle$ with e inducing an outer automorphism on X and M.

Proof. Let Y of order 5 in F with $E(C_F(Y)) = B \cong F_5$; we may choose our non-2-central involution t to invert Y and centralize B. By 15.5 there is an A_{12}-subgroup K of B and by 15.6 K centralizes a conjugate $s \neq t$ of t in H. Let $W = \langle Y,t,s \rangle$ so that $W \leq C(K)$. By 15.4 and its proof and by 15.5, there is $M_{12} \cong I \leq K$ such that $C_F(I) = J \cong L_2(11)$. Then $W \leq J$ so $W \cong A_5$ or $W = J$. The latter is impossible as an element of order 11 centralizes no A_{12}-subgroup. Indeed this argument shows

$W = C_F(K)$. From the structure of $N(Y)$ described in [23], $N_F(Y) \cap N(K) = YK\langle v\rangle$ where $v^2 = t$ and v induces an outer automorphism on K. Hence $N_F(K)$ is as described in 15.7.

Now let $X \times M$ satisfy the hypothesis of the lemma. Then a subgroup of X of order 5 is conjugate to Y in F, (From [23], members of the second class of Z_5-subgroups centralize no A_{12}-subgroup.) so without loss $Y \leq X$. Further K is determined up to conjugacy in B by its isomorphism type so without loss $M = K$. But then $X = C_F(M) = W$. It remains only to show X is a root A_5-subgroup of F and $N_F(X) = N_F(M)$. But setting $U = N_X(\langle t,s\rangle)$, we saw in paragraph 1 of this section that $N_F(U) = LU$, where $L \cong O_{10}^-(2)$ so in particular X does not centralize B as $F_5 \not\leq \Omega_{10}^-(2)$. Hence as M is maximal in B, $N_F(X) = N_F(M)$. Moreover there is a conjugate U^g of U in M with $C_M(U^g) = D^g \cong A_8$. Now $X^{g^{-1}} \times D \leq L$; we conclude $X^{g^{-1}}D$ stabilizes a nondegenerate subspace of dimension 4 and sign $-$ of the orthogonal space for L. It follows from 15.1 that X is a root A_5-subgroup of F.

(15.8) F_2 contains a unique class of $\Omega_{10}^-(2)$ subgroups.

Proof. Existence follows from 15.1. On the other hand if L is such a subgroup of $G = F_2$ then L contains an A_{12}-subgroup A. Let Δ consist of those A_5-subgroups of A moving 5 of the 12 points permuted by A. By 15.6 each X in Δ is a root A_5-subgroup of G and as $E(C_L(X)) \cong A_8$, $E(C_L(X)) = E(C_G(X))$ by 15.1. Hence $L = \langle E(C_G(X)) : X \in \Delta\rangle$ so as A is determined up to conjugacy in G by 15.6, the proof is complete.

(15.9) Let $M = M(24)$ and $G = [M,M]$. Let $\Gamma = \Gamma(\mathfrak{A},\mathfrak{A},G)$, and Γ^* the set of maximal members of Γ under inclusion. Then

(1) There are two orbits of G on Γ^* with representatives $L_i = L_{i1} \times L_{i2}$, for $i = 1,2$. The normalizer of each normal subgroup of each member of Γ normalizes some member of Γ^* containing it.

(2) $N_M(L_1) \cong S_5 \times S_9$ is a D-subgroup of M and $N_M(L_{1i}) = N_M(L_1)$ for $i = 1,2$.

(3) $N_M(L_2) \cong S_3 \times (S_6 wr Z_2)$ is a D-subgroup of M with $N_G(L_{2i}) \leq N_G(L_2)$ for $i = 1,2$.

(4) Up to conjugacy G possesses a unique A_{12}-subgroup K. $N_M(K) \cong S_{12} \times Z_2$ is a D-subgroup of M.

(5) Up to conjugacy G possesses a unique $\Omega_{10}^-(2)$-subgroup I. $N_M(I) \cong O_{10}^-(2)$ is a

D-subgroup of M.

Proof. By 15.2 there is a D-subgroup I of M isomorphic to $O_{10}^{-}(2)$ and of course I contains a D-subgroup K isomorphic to S_{12}. By 15.7 there is $XX_0\langle e\rangle \leq F$ with X a root A_5-subgroup, $X_0 \cong A_{12}$, and e inducing an outer automorphism on X and X_0. There is $Z_3 \cong U \leq X_0$ normalized by a subgroup L_1 of $XX_0\langle e\rangle$ isomorphic to $S_5 \times S_9$ and $N_F(U)/U \cong M(24)$ so the existence of a D-subgroup L_1 as in (2) with $X = L_{11}$ and with L_{12} the S_9 factor of L_1 is also established. By construction $L_1 = N_M(X)$.

Let V be of order 5 in G; then $N_M(V) \cong F_{20} \times S_9$, where F_{20} is the Frobenius group of order 20. It follows from 7.6 that the simple subgroups in $\mathfrak{A}$ centralizing V are isomorphic to A_n, $5 \leq n \leq 9$, and that up to conjugacy there is a unique such subgroup G_n unless n = 5, where there are two such groups. In particular we may take $V \leq X$, forcing $L_{12} = E(C_G(V))$ and showing L_1 is determined up to conjugacy in G by its isomorphism type.

There is W of order 7 in G_7 and as X centralizes W, $N_M(W) \cong F_{42} \times S_7$ from [23]. Hence there is a 3-transposition d centralizing WX with $N_{C(d)}(X) \cap N(W)$ maximal in N(W) so either $X = E(C_G(G_7))$ or G_7 centralizes $E(C_G(W))$. The later is impossible as an element of order 7 in $E(C_G(W))$ is 7-central in G and hence has a solvable centralizer from [23].

This shows that if $E \times D \in \Gamma$ then $E \cong A_n$, $D \cong A_m$, and if $n \geq 7$ then ED is contained in a conjugate of L_1. Moreover to complete the proof of (1)-(3) it remains to consider the case where n = 5 or 6 and in the former case E is not a conjugate of X. In either event we may take $V \leq D$ and E contains a subgroup Z of order 3 moving 6 of the 9 points moved by G_9. Thus working in S_9 where $D \cap S_9$ is the set of ordinary transpositions, we see $Z = \langle z_1 z_2\rangle$ where z_i is inverted by a 3-transposition of S_9 centralizing z_{3-i}. Then $C_M(Z) = Z \times Q$ where $Q \cong Z_2/\Omega_6^{-}(3)/E_{3^6}$ is generated by 3-transpositions. Next EX centralizes a D-subgroup $\langle d_3, z_3\rangle = B$ of $N_M(L_1)$, where $d_3 \in D$ inverts z_3 of order 3. As $[B,X] = 1$, $z_3 \in O_3(Q) = P$. Further PZ is an orthogonal space for $N_M(PZ)/PZ \cong Z_2/\Omega_7(3)$ (cf. 29.10) from which it can be seen that there is a N(PZ)-conjugate Z' of Z in each nontrivial X-submodule $P_0 = [P_0,X]$ of P. Hence if $[C_P(E),X] \neq 1$ then $E \leq C(Z')$ contradicting $Z \leq O_3(C(Z'))$. Further E centralizes an S_6-subgroup D_0 of K containing X and as $X \leq C(C_P(E))$, $D_0 = [D_0,X] \leq C(C_P(E))$. So D_0 centralizes B. Finally $C_M(BED_0) = 1$ from which it follows that $B \times D_0 = C_M(E)$. This completes the proof of (1)-(3).

It remains to establish the uniqueness statements in (4) and (5) so let J be a subgroup of G satisfying the hypothesis of (4) or (5). Then J possesses a subgroup $X' \times E \cong A_5 \times A_n$, n = 7 or 8, so by (1)-(3) we may take X' = X and E is determined up to conjugacy in $Y = E(C_G(X))$. Further for each A_6-subgroup D of E, $E(C_J(D)) \cong A_6$, so by (1)-(3), $E(C_J(D)) = E(C_G(D))$. Finally $J = \langle E, E(C_J(D))\rangle$, so J is determined up to conjugacy in G.

(15.10) Let G = M(23) or M(22) and $\Gamma = \Gamma(\mathfrak{A},\mathfrak{A},G)$. Then

(1) Up to conjugacy there exists a unique subgroup L of G isomorphic to A_{12} or $\Omega_8^+(2)$ for G = M(23) or M(22), respectively.

(2) $N_G(L)$ is a D-subgroup of G containing the normalizer of a conjugate of every member of Γ. $N_G(L) \cong S_{12}$ or $Aut(\Omega_8^+(2))$, respectively. If G is M(22) then $N_G(L)$ is strongly 5-embedded in G; indeed $N_G(L) = \Gamma_{1,T}(G)$ for $T \cong Syl_5(G)$.

(3) If G is M(23) then up to conjugacy it possesses a unique $Sp_8(2)$-subgroup K. K is a selfnormalizing D-subgroup.

Proof. When G is M(23) the proof of parts (1) and (2) is similar to that of the previous lemma except much easier. Suppose G is M(22). We can regard G as $I/\langle u\rangle$ where u = ab is the product of 3-transpositions a and b in M = M(24) and $I = E(C_M(u))$. Now if d is a 3-transposition commuting with neither a nor b then $S = \langle a,b,d\rangle \cong S_4$ and $C_M(S) \cong Aut(\Omega_8^+(2))$. (cf. 13.6.5 in [13]) Hence G contains a subgroup $B \cong Aut(\Omega_8^+(2))$. Let $T \in Syl_5(B)$. Then $N_G(X) \cong N_B(X)$ for each $1 \neq X \leq T$ and $B = \Gamma_{1,T}(B)$ so $B = \Gamma_{1,T}(G)$ is strongly 5-embedded in G. Thus (1) and (2) hold in this case too.

Finally let G = M(23). The existence of an $Sp_8(2)$-subgroup of G follows from 15.2. Conversely let K be such a subgroup and S an S_4-subgroup of K generated by transvections. Then $C_K(S)) = C_K(S\langle d\rangle) \cong Sp_4(2)/E_{32}$, where d is the transvection whose center is the radical of the commutator space of S on the symplectic space of K. The usual argument shows d is a 3-transposition and as $C_G(d) \cong M(22)/Z_2$, we find $C_G(S\langle d\rangle) \cong C_K(S\langle d\rangle)$ and hence conclude the groups are equal. Now let U of order 5 in K and Y the subgroup of $C_K(U)$ generated by its transvections. Then $Y \cong S_6$ and we may take $S\langle d\rangle \leq Y$. As $\langle C_D(U)\rangle \cong S_7$, Y is determined up to conjugacy in $C_G(U)$ so $K = \langle Y, C_G(S\langle d\rangle)\rangle$ is determined up to conjugacy in G.

(15.11) Let X be a root A_5-subgroup of F and $Y = Y_1$ an A_6-subgroup of $C_F(X)$ with $C_F(XY) = Y_2 \cong A_6$. Then

(1) $E(C_F(Y)) = Y_2 \times Y_3$ with $Y_3 \cong A_6$.

(2) Let $I = Y_1 \times Y_2 \times Y_3$. Then $N_F(I)/I \cong D_8/Z_3$ and $N_F(I)$ is transitive on $\{Y_1,Y_2,Y_3\}$.

(3) $N_F(Y_2Y_3) = Y(N_F(Y_2Y_3) \cap N(X))$, so $Aut_F(Y) \cong S_6$.

Proof. $E(N_F(X)) = L \cong A_{12}$ by 15.7. Hence Y is determined up to conjugacy in L by its isomorphism type and that of Y_2. Let $Z_3 \cong U \leq X$. $C_F(U)/U \cong M(24)'$ so by 15.9, $C_F(UYY_2) = V \cong Z_3$. Now $Y_3 = \langle V,X\rangle \leq C_F(YY_2) \leq C_F(Y_2X_1)$, where X_1 is a root A_5-subgroup of Y. So $Y_3 \cong A_6$. Let $L_1 = C_F(X_1)$. $N_{L_1}(Y_2Y_3)$ is maximal in L_1 by 7.4, so as $X_1 = C_F(L_1)$ it follows that $Y_2Y_3 = E(C_F(Y))$. Thus (1) is established.

Let W be of order 5 in Y. Then $N_F(W) \cong Aut(F_5)/D_{10}$ by [23]. Using 15.12 a little early, $(N_F(I) \cap N_F(W))/N_Y(W)$ is contained in an S_{12}-subgroup of $N_F(W)/N_Y(W)$. In particular $N_F(W) \cap N_F(I)$ is transitive on $\{Y_2,Y_3\}$, and $(N_F(W) \cap N_F(I))/N_I(W) \cong D_8$, from which (2) follows. Also by (2) and 15.7, $N_F(Y_2Y_3) = Y(N_F(Y_2Y_3) \cap N(X))$, so (3) holds.

(15.12) Let $G = F_5$, $A = Aut(G)$, and $\Gamma = \Gamma(\mathfrak{A},\mathfrak{A},G)$. Then

(1) Up to conjugacy in G there is a unique subgroup L of G isomorphic to A_{12}.

(2) $N_A(L) \cong S_{12}$ and contains a conjugate of the normalizer of each normal subgroup of each member of Γ.

Proof. L exists by 15.5. Indeed we saw in the proof of 15.5 that for any such L, if U is of order 3 in L moving 3 of the 12 points permuted by L, then $N_G(U) = N_L(U)$. But then $L = \langle N_G(U), N_G(U_1)\rangle$, where U_1 moves 3 of the 9 points permuted by $E(N_G(U))$, so as U is determined up to conjugacy in G, so is L. Moreover all elements of G of order 3 with $C_G(u)$ of 2-rank at least 2 are fused into U by [23], so if $A \times B \in \Gamma$ then we may take $U \leq A$ and hence by 7.6 B is an alternating group essentially determined up to conjugacy in $C(U)$ by its isomorphism type. Next if V is of order 7 in G then $C_G(V)^{\infty} \cong A_5$. This shows that if $B \cong A_n$ for $n \geq 7$ then A is conjugate to $E(N_G(V))$. Now we may take $G = E(N_F(W))$ for some W of order 5 in F and indeed take $W \leq C_F(X)$ for some root A_5-subgroup X of F. Then $E(N_G(X)) \cong A_5 \times A_7$ which together with the last remark imply $N_A(X)$ is contained in some conjugate of $N(L)$ and handles the case $n \geq 7$. So we may assume $B \cong A_6$ or A_5 and if $B \cong A_5$ then $B \notin X^G$. Then $D = E(C_L(B))$ $\cong A_6$. If $B \cong A_6$ let $U_1 \times U_2 \in Syl_3(B)$ with $U_i \cong Z_3$. Then $DU_1U_2 \leq C_G(U_1) \cong A_9$, so

$D = C_G(U_1U_2) \geq C_G(B)$. Hence we may assume $B \cong A_5$ but $B \notin X^G$. If W is of order 5 in B then as $[D,W] = 1$, $D \leq E(C_G(W)) = E \cong U_3(5)$. Then as B centralizes no element of order 7 it follows that $D = E(C_G(B))$ and the proof is complete.

(15.13) Let $\Gamma = \Gamma(\mathfrak{A},\mathfrak{A},\mathfrak{A},F)$. Then

(1) Up to conjugation there are two maximal members L and K of Γ.

(2) $L \cong A_6 \times A_6 \times A_6$ and $N_F(L)/L \cong D_{8/S_3}$.

(3) $K \cong A_5 \times A_5 \times A_7$ and $N_F(K)/K \cong D_8$. The A_5 factors of K are root A_5-subgroups of G.

(4) Each A_6 factor of L has two classes of A_5-subgroups and these classes are not fused in F.

(5) The normalizer of each member of Γ normalizes some conjugate of L or K.

Proof. The existence of L and K follows from 15.7 and 15.11.

Let J be an A_6 factor of L. Then by 7.6, J has two classes of A_5-subgroups. Let X and X′ be representatives of these classes; we may choose X to be a root A_5. Hence $C_L(X) = P$ is determined up to conjugacy in $C_F(X)$ by its isomorphism type so if X′ is fused to X in F it is even fused in $N_F(P)$. But by 15.11.3, $N_F(P) = J(N_F(P) \cap N(X))$, a contradiction. Hence (4) is established.

Let $A \times B \times C \in \Gamma$ and let U of order 5 in A. It's easy to see $\Gamma(\mathfrak{A},\mathfrak{A},J_2)$ is empty (cf. 16.4) so $E(C_F(U)) = E \cong F_5$. Hence by 15.12 and 15.7, $BC \leq Y = C_F(X) \cong A_{12}$ for some root A_5-subgroup X of F. Hence $B \cong A_n$ and $C \cong A_m$, with $m+n \leq 12$, and by symmetry $A \cong A_k$, with $k \leq 12-n$ and $12-m$. Moreover $\langle X,A\rangle = A' \cong A_{k'}$. Now if 2 of m,n,k are at least 6 then by 15.11, ABC is contained in a conjugate L′ of L and $N(ABC) \leq N_F(L')$. So assume two of m,n,k are 5. If some factor is a root A_5, then by 15.7, ABC is contained in some conjugate K′ of K and by 15.7 and (4), $N(ABC) \leq N(K')$. Thus we may assume $m = k = 5$ and no factor is a root A_5. Replace A by A′ and induct on k to get $ABC \leq L'$ and $N(A'BC) \leq N(L')$. If $n = 6$ then $N(ABC) \leq N(C) \leq N(L')$ by 15.11, so we can take $n = 5$. Now $N(ABC) \cap N(L)$ is transitive on $\{A,B,C\}$ by 15.11, as none is a root A_5. Also $N(A) \cap N(BC) \leq N(ABC)$ by 15.11 and 15.12, so the proof is complete.

(15.14) Let G = He and $\Gamma = \Gamma(\mathfrak{A},\mathfrak{A},G)$. Then up to conjugacy Γ contains a unique member L. $L \cong A_5 \times A_5$ and $N_G(L)/L \cong E_4$.

Proof. We may take $G = E(N_F(U))$ for a suitable subgroup U of F of order 7. Then by 15.13 G contains an $A_5 \times A_5$ subgroup, after which the rest is easy since $E(C_G(V)) \cong A_5$ for V of order 5 in G.

(15.15) F contains no $Sp_{10}(2)$ subgroup.

Proof. Assume $Sp_{10}(2) \cong L \leq F$ and let $S_3 \cong S \leq L$ be generated by transvections. Then $K = C_L(S) \cong Sp_8(2)$ so as usual using [23], $C_F(S) \cong M(23)$. Hence by 15.10 K is determined up to conjugacy in $C_F(S)$ and then by 15.2, $K \leq M \cong \Omega_{10}^-(2)$, with $M \leq C_F(U) \cong M(24)'/Z_3$ and $U = O_3(S)$. Now K contains a subgroup $Y = Y_1 \times Y_2$ with $Y_i \cong A_6$ and $C_L(Y_i) \cong Sp_6(2)$. By 15.9.1, Y is determined up to conjugacy in $C_F(U)$. Now U is contained in a root A_5-subgroup X of F. By 5.7, $I = C_F(X) \cong A_{12}$. Thus as Y is determined up to conjugacy in $C_F(U)$ and I contains a subgroup isomorphic to Y, we may take $Y \leq I$. But now by 15.11, $C_F(Y_i)^\infty \cong A_6 \times A_6$, contradicting $C_L(Y_i) \cong Sp_6(2)$.

(15.16) (1) F_2 has a unique class of F_3-subgroups.

(2) The normalizers of each ${}^3D_4(2)$-subgroup and of each $F_4(2)$-subgroup of F_2 centralize involutions.

(3) F_3 has a unique class of ${}^3D_4(2)$-subgroups; the normalizer of such a subgroup is $Aut({}^3D_4(2))$.

(4) $M(24)'$ has two classes of He-subgroups fused in M(24). The normalizer in $M(24)'$ of such a subgroup is Aut(He).

Proof. Let W be of order 3 in F with $N_F(W) \cong Z_3 \times F_3$. The involution inverting W and centralizing E(C(W)) has centralizer in F isomorphic to F_2/Z_2, which establishes the existence of an F_3-subgroup of F_2. Similarly there is Z of order 7 in F with $C_F(Z)) \cong Z_2/(He \times F_{21})$ so an element x of order 3 faithful on Z and centralizing E(C(Z)) has $N_F(\langle x\rangle)/\langle x\rangle \cong M(24)$, establishing the existence of an Aut(He)-subgroup of $M(24)'$. The proof that there are exactly two classes fused in M(24) is left until section 21.

Let T be a Sylow 7-group of $G \cong F_2$ and $Q = O^2(N_G(T))$. Let U be of order 7 in T. We saw in paragraph 2 of the proof of 15.3 that there are just 3 $L_3(2)$-subgroups A_i, $1 \leq i \leq 3$, of $E(C_G(U))$ containing $T \cap E(C_G(U))$, and that $C_G(UA_1) = UE$, $E \cong E_4$, $C_G(E) \cong$

$E_4 \times F_4(2)$, and A_2 is fused to A_3 in E. Let $I_i = \langle A_i, Q\rangle$. T is contained in an $Aut(^3D_4(2))$ subgroup I of $F_4(2) \cong E(C_G(E)) = Y$ with $E(C_I(U)) \cong L_3(2)$, $Q \cong O^2(N_I(T))$, and $I = \langle E(C_I(U)), O^2(N_I(T))\rangle$. So $I = I_1$. Also A_1 is fused to A_2 in $Aut(O^2(C_G(s))/\langle s\rangle)$ where $\langle s\rangle = Z(E(C_G(U)))$, since a diagonal automorphism of $O^2(C_G(s))/\langle s\rangle \cong {}^2E_6(2)$ induces a diagonal automorphism of $E(C_G(U))/\langle s\rangle \cong L_3(4)$ transitive on the 3 classes of $L_3(2)$-subfgroups. So $I_i \cong I$ for each i. This completes the proof of the first part of (2), since any $^3D_4(2)$-subgroup of G containing T is I_i or I_i^g where $N_G(T) = \langle g\rangle QE$. Next $\langle I_1, g\rangle E = N_G(E)$ or else the former group is strongly 7-embedded in the latter contrary to [34]. So as $F_4(2)$ is not a section of F_3 by [23], while $N_G(T) = N_L(T)E$ and $E(C_L(U)) \cong L_3(2)$ for any F_3-subgroup L of G containing T, it follows that I_i and I_i^g, i = 2 or 3 and $g \in N_L(T)-C(E)$, are the $Aut(^3D_4(2))$-subgroups of L containg T. This establishes (3). Further we will see in Theorem B of section 20 that I_2 is maximal in L, so $L = \langle I_i, I_i^g\rangle$ is determined up to conjugacy in G, completing the proof.

(15.17) Aut(M(22)) has a $^2F_4(2)$-subgroup.

Proof. Let G be F_2 and d,e be 3,4-transpositions of G with $C_G(\langle d,e\rangle) \cong E_4 \times F_4(2)$. There exists a conjugate a of d with $V = \langle a,d,e\rangle \cong S_4$. Let $U = O_2(V)$ and $u \in U-\langle ed\rangle$. Then $C_G(ed) = C_G(\langle e,d\rangle)\langle u\rangle$ with u inducing an outer automorphism on $E(C_G(\langle e,d\rangle))$, so $C_G(U) = U \times X$ with $X \cong {}^2F_4(2)$. By symmetry among the members of d^V, $[V,E(X)] = 1$, so E(X) centralizes $\langle a,d\rangle$. However $C_G(\langle a,d\rangle) \cong Aut(M(22))$, so the lemma holds.

16. Groups with abelian Sylow groups.

In this section assume G is a sporadic simple group, p is a prime, and P is a noncyclic abelian Sylow p-subgroup of G.

(16.1) One of the following holds:

(1) $P \cong E_8$ and G is J_1.

(2) $P \cong E_9$ and G is M_{11}, M_{22}, M_{23}, or HS.

(3) $P \cong E_{25}$ and G is J_2, Sz, He, M(22), M(23), or M(24)'.

(4) $P \cong E_{49}$ and G is Co_1, F_3, or F_2.

(5) $P \cong E_{121}$ and G is F_1.

(6) $P \cong E_{81}$ and G is ON.

Proof. See [23] for example.

I take the point of view that the subgroup structure of the Mathieu groups is best studied from the point of view of their action on the Steiner system S(5,8,24) and appeal to the list of maximal subgroups in [19]. Thus in the remainder of this section I assume G is not a Mathieu group, unless the restriction is explicitly relaxed.

(16.2) The conjugacy classes X^G of subgroups of order p in G and $F^*(N_G(X))$ are listed in Table 16.2. If there are two classes, then half the subgroups of P of order p are in each class.

Again this information can be found in [23]. The next result follows from 2.2 together with 16.2 and occasionally some extra information about the centralizers of elements of order p such as in [23].

TABLE 16.2

G	J_1	HS	He	Sz
$F^*(N_G(X))$	$Z_2 \times A_5$	$Z_3 \times A_5$	$Z_5 \times A_5$	$Z_5 \times A_6$, $Z_5 \times A_5$

G	J_2	$M(22)$	$M(23)$	$M(24)'$
$F^*(N_G(X))$	$Z_5 \times A_5$, E_{25}	$Z_5 \times A_5$	$Z_5 \times A_7$	$Z_5 \times A_9$

G	ON	F_1
$F^*(N_G(X))$	$E_9 \times A_6$	$Z_{11} \times M_{12}$

G	F_3	F_2	Co_1
$F^*(N_G(X))$	$Z_7 \times L_3(2)$	$Z_7 \times L_3(4)/Z_2$	$Z_7 \times L_3(2)$, $Z_7 \times A_7$

(16.3) One of the following holds:

(1) $\mathcal{N}_G(P,p') = 1$.

(2) G is $M(23)$, $M(24)'$, or F_2 and $\langle \mathcal{N}_G(P,p') \rangle = O^p(C_G(P)) \cong Z_2$, A_4, or E_4, respectively.

(3) G is HS and each member of $\mathcal{N}_G(P,3')$ is a 2-group.

Let $\mathfrak{A}$ be the set of nonabelian simple groups of order divisible by p and $\Gamma = \Gamma(\mathfrak{A},\mathfrak{A},G)$ the set of subgroups $A \times B$ of G with A and B in $\mathfrak{A}$.

(16.4) If Γ is nonempty then G is not J_1, J_2, HS, ON, or F_3.

Proof. Assume to the contrary that G is one of these groups and $A_1 \times A_2 \in \Gamma$. Let X_i of order p in A_i; then $A_{3-i} \le E(C_G(X_i)) = L_i$.

If G is ON then $L_i \cong A_6$ so $A_i \cong A_5$ or A_6. This is impossible as G has cyclic Sylow 5-groups while A_1A_2 does not.

If G is J_1 then $|G|_2 = 8 < |A_1A_2|_2$, a contradiction.

If G is J_2 then a subgroup Y_i of order 3 in L_i is 3-central in G, so as the involutions in $E(C_G(Y_i))$ are 2-central, the same holds for those in L_i. Hence if z is an involution in L_2 then $\langle z\rangle$ is a Sylow 2-group of $C_G(X_1\langle z\rangle)$, since this holds for each subgroup X_1 of order 5 centralizing a 2-central involution z. But of course $\langle z\rangle$ is not even Sylow in the centralizer in L_2 of $\langle z\rangle X_1$.

If G is HS let Y_i be of order 5 in A_i. Then $C_G(Y_i)$ is nonsolvable so Y_i is not 5-central in G. Thus $E(C_G(Y_i)) \cong A_5 \cong L_i$, so $L_i = E(C_G(Y_i))$. But an element of order 5 in $E(C_G(Y_i))$ is 5-central, a contradiction.

This leaves $G = F_3$. Then there exists an involution t inverting X_i with $L_i\langle t\rangle \cong PGL_2(7)$. Now t acts on $E(C_G(L_i)) = A_i$ and then on an involution z in A_i. So $PGL_2(7)$ is involved in $C_G(z)$, contradicting $C_G(z)/O_2(C_G(z)) \cong A_9$.

(16.5) If G is Co_1 or Sz then up to conjugation Γ contains a unique maximal member $L_1 \times L_2$. $L_{3-i} = E(C_G(x_{3-i})) = C_G(L_i)$, where x_i is an element of order p in L_i and x_1 and x_2 are representatives for the two classes of elements of order p in G.

Proof. Suppose first that G is Co_1. Then G contains a TI-set $E \cong E_4$ with $N_G(E) \cong Z_2/(A_4 \times G_2(4))$. Let U be of order 3 in $E(C_G(E)) = L$ with $E(C_L(U)) \cong SL_3(4)$; there is an involution s fused into E in G inverting U, inducing a field automorphism on L and with $\langle s\rangle C_G(L) \cong S_4$. Then $K = C_L(U\langle s\rangle) \cong L_3(2)$ and $K \leq E(C_G(s)) = I$ so $Y = \langle C_G(L), C_L(I), U\rangle \leq C(K)$. Let x_1 of order 7 in K. Then $C_G(x_1) \cong Z_7 \times A_7$ from which it follows that $Y = E(C_G(x_1))$. Further if x_2 is of order 7 in Y then $C_G(x_2) \cong Z_7 \times L_3(2)$ so the proof is complete in this case. Incidentally if V is of order 3 in $C_G(L)$ then $C_G(V) \cong Sz/Z_3$ so Sz has a $G_2(4)$-subgroup, a fact we will need in a moment.

Next let G be Sz and H a $G_2(4)$-subgroup of G. Then H has an $A_5 \times A_5$ subgroup $A \times B$. let x and y of order 5 in A and B, respectively. Then x and y are representatives for the classes of elements of order 5 in G and we may take $A = E(C_G(y))$. Next from 11.1.6 there is a TI-set $E \cong E_9$ with $C_G(E) \cong A_6 \times E_9$. Let x' of order 5 in $C_G(E)$; as $m_3(C_G(x')) > 1$ and $C_G(x) \cong Z_5 \times A_6$, we may take $x = x'$ and $E \cap B \neq 1$. Then $A \leq C(E \cap B) \leq C(E)$ so $I = \langle B,E\rangle \leq C(B) \leq C(x) \cong Z_5 \times A_6$, from which the lemma follows.

(16.6) Up to conjugacy, G = He contains a unique subgroup L isomorphic to $Sp_4(4)$. $|N_G(L):L| = 2$.

Proof. Assume G is He and $Sp_4(4) \cong L \le G$. Let $T \in Syl_2(L)$. Then $T = A_1A_2$ with $A_i \cong E_{64}$ and $V = A_1 \cap A_2 \cong E_{16}$. But if $T \le S \in Syl_2(G)$ then $J(S) \cong T$ so $T = J(S)$. Hence from the 2-structure of He in [23] (or [14], section 4 for more details), $V = E_1 \times E_2$, where $C_G(E_i) \cong L_3(4)/E_4$, and where all elements of V not in E_1 or E_2 are 2-central in G. On the otherhand from the structure of L, $N_L(T) = T(Y_1 \times Y_2)$ with $Y_i \cong Z_3$ and $V = V_1 \times V_2$ where $V_i = C_V(Y_i)$. Hence $V_i = E_i$. Next A_1 or A_2 is normal in $N_L(V_i)$, so as V_1 is fused to V_2 in $N_G(A_1) \cap N_G(A_2)$, we may take $A_i \trianglelefteq N_L(V_i)$. Now setting $L_i = N_L(V_i)^{\infty}$, $L_i/V_i \cong L_2(4)$. Thus $L_i = K_i$, where $K_i = (N_G(A_i) \cap N_G(V_i))^{\infty}$. So $L = \langle L_1, L_2 \rangle$ is determined up to conjugacy in G. Further $N_G(K_1) = N_L(V_1)\langle x \rangle$, where x induces a field automorphism on L_1/V_1 and acts on K_2. Thus x induces an outer automorphism on L and $N_G(L) = L\langle x \rangle$.

It remains to demonstrate the existence of an $Sp_4(4)$-subgroup of G. First by 15.14 there is $X_1 \times X_1^x \le G$ with $X_1 \cong L_2(4)$ and x an involution. As a Sylow 2-group of X_1 centralizes an element of order 5, we may take $V_1 \in Syl_2(X_1^x)$. Then X_1V_1/V_1 is generated by a pair of root groups in $C_G(V_1)/V_1$ and hence is determined up to conjugacy. So we may take $K_1 = X_1A_1$.

Also without loss $A_2 = V_1^x C_{A_1}(V_1^x)$. Then $A_2 \le C_G(V_1V_1^x) = S$. Of course S is x-invariant and S is a Sylow 2-group of $C_G(V_1)$ with $\mathfrak{A}(S) = \{A_2, A_1'\}$. As $A_1' \notin (A_2)^G$, we conclude $x \in N_G(A_2)$.

Let $K = \langle K_1, x \rangle$, $\mathcal{P}_1 = (V_1)^K$, $l = l(V_1, V_1^x) = (V_1)^G \cap A_2 = \{V_1\} \cup (V_1^x)^{A_1}$, and $\mathcal{L}_1 = l^K$. I'll show $\mathcal{S}_1 = (\mathcal{P}_1, \mathcal{L}_1)$ is the generalized 4-gon of $Sp_4(4)$ and V_1 is the root group stabilizing the set $\Delta(V_1) = \{V_1^{xk} : k \in K_1\}$ of points colinear with V_1 in $\mathcal{S}_1$. From this it follows that $K^{\infty} \cong Sp_4(4)$.

Let $\Sigma(V_1) = \bigcup_{D \in \Delta(V_1)} (\Delta(D) - l(V_1, D))$. Claim K is transitive on $\Sigma(V_1)$ and $X_1 \cap \mathcal{P}_1 = \Delta(V_1) \cap \Delta(V_1^y)$ for $y \in X_1^x - K_1$. First $X_1 \cap \mathcal{P}_1 \subseteq \Delta(V_1)$ and $y \in X_1^x \le C_K(X_1)$, so $X_1 \cap \mathcal{P}_1 \subseteq \Delta(V_1) \cap \Delta(V_1^y)$. Further $\Delta(V_1) \cap \Delta(V_1^y) \le C(\langle V_1, V_1^y \rangle) = C(X_1^x) = X_1$, so we have equality. Next K_1 is transitive on $\Delta(V_1)$, so it remains to show $N_{A_1}(D)$ is transitive on $\Delta(D) - l$, where $D = V_1^x$. But $N_{A_1}(D) = A_1 \cap A_2 = V$, so as $x \in N(A_2)$, $V \le K_1^x$. Indeed V_1 is regular on the lines through D distinct from l and $V \cap A_1^x$ is regular on k - {D} for each such line k as $V_1^x(V \cap A_1^x) = C_{A_1^x}(V_1)$. So the claim is established.

Hence as X_1 contains a member of $\mathcal{P}_1$ in each member of $\mathcal{L}_1$ through V_1, the claims and a standard argument shows $\mathcal{P}_1 = \{V_1\} \cup \Delta(V_1) \cup \Sigma(V_1)$ and $\mathcal{S}_1$ is a generalized 4-gon. We then easily calculate the order of $\mathcal{P}_1$ and $\mathcal{L}_1$, after which another standard argument shows $\mathcal{S}_1$ is the 4-gon for $Sp_4(4)$.

The next few lemmas discuss the Fischer groups M(24), M(23), and M(22), each of which is generated by a set D of 3-transpostions. Section 13 in [13] contains a discussion of these groups and facts about 3-transposition groups which the reader may wish to inspect.

(16.7) M(24) has no $U_3(4)$-subgroup.

Proof. Suppose M is M(24) and $U_3(4) \cong L \le M$. Let $T \in Syl_2(L)$, $Z = Z(T)$, and X of order 5 in $N_L(T)$. There exists Y of order 5 in $C_L(Z)$ and $Z \le K = E(C_L(Y)) \cong L_2(4)$. Now $A = C_M(Y) \cong S_9$ with $A \cap D$ the set of transpositions in A. So K moves 5 or 6 of the 9 points permuted by A. Hence $z \in Z^\#$ moves 4 of these points so $z = ab$ where $a,b \in D$ and D is the set of 3-transpositions of M. Indeed either Z moves 4 points and $Z \le S = \langle S \cap D \rangle \cong S_4$ or Z moves 6 points and $Z \le \langle a,b,d \rangle = E \cong E_8$, for some $d \in D$. In any case $U = XT \le O^2(C_G(Z)) \le C(\langle a,b \rangle)$, so U centralizes S or E, respectively. However in the first case $Z = Z \cap U \le S \cap C_M(S) = 1$, a contradiction so the second holds. Hence $U \le I = O_2(C_G(E)) \cong U_6(2)/E_4$. Now $E_{16} \cong T/Z \le I/Z \cong U_6(2)$, impossible as $Z = \Omega_1(T)$ while every involution in I/Z lifts to an involution in I. (cf. 21.7 in [12]).

(16.8) (1) Every A_{10} and A_{11} subgroup of M(24) and every A_{10} subgroup of M(23) centralizes a 3-transposition of the respective group.

(2) Up to conjugacy there is a unique A_{11}-subgroup L of G = M(23). $N_G(L) \cong S_{11}$ is a D-subgroup of G.

(3) Up to conjugacy there are two A_{10}-subgroups L_i of G = M(22). L_1 is fused to L_2 in Aut(G) and $N_G(L_i) \cong S_{10}$ is a D-subgroup.

(4) Each $P\Omega_8^+(3)$-subgroup of M(24) centralizes a D-subgroup isomorphic to S_3.

(5) Up to conjugacy there is a unique $P\Omega_8^+(3)$-subgroup L of G = M(23). $N_G(L)/L \cong S_3$ and $N_G(L)$ is a D-subgroup.

Proof. Let $M = M(24)$ and $S_3 \cong S$ a D-subgroup of M. Then $C_M(S) \cong S_3/P\Omega_8^+(3)$ is a D-subgroup, supplying the existence of such a subgroup in M and in M(23). Next if $P\Omega_8^+(3) \cong L \le G \cong M(24)'$ or M(23) then considering the action of L on its natural module, there exist 3 subgroups I_i of L containing a Sylow 5-group P of L and with $I_i = I_{i1} \times I_{i2}$ where $I_{ij} \cong A_6$. (ie. I_{ij} centralizes a 4-dimensional subspace of sign − of the natural module.) Let $P_{ij} = P \cap I_{ij}$; notice $(P_{ij} : i,j)$ is the set of subgroups of P of order 5. Let $Q = P_{21}$. Then $C_G(Q) \cong A_9$ or S_7, for $G \cong M(24)'$ or M(23), respectively, so the intersection of all A_6-subgroups of $C_G(Q)$ containing P is $J \cong A_5$. Now as $L = \langle I_1, J\rangle$ and I_1 is determined up to conjugacy in G by 15.9 and 15.10, it follows that L is determined up to conjugacy in G. Using 15.10 it is not difficult to see $N_G(L)$ is as claimed when G is M(23).

Similarly alternating subgroups of the prescribed type exist by 15.9 and 15.10. So assume $A_n \cong L \le G$ with n = 10 or 11 and $G \cong M(24)'$, M(23), or M(22). Let $P \in Syl_5(L)$; again there is $I = A \times B \le L$ with $P \le I$, $A \cong A_5$, and $B \cong A_{n-5}$. By 15.9 and 15.10, I is contained in a D-subgroup $K \cong S_m$, where m = 10 if G is M(22) and m = 12 otherwise, and such that A moves 5 or 6 of the set Δ of m points moved by K. Let Z be of order 3 in A.

Suppose first that A moves 6 points of Δ. Then $C_G(Z) \cong Z_3 \times (\Omega_6^-(3)/E_{3^6}$, $Z_3 \times (Z_2/\Omega_5(3))/E_{3^5}$, or is a 2,3-group for $G = M(24)'$, M(23), or M(22), respectively. So as $A_{n-3} \cong C_L(Z)$ it follows that $G = M(24)'$ and n = 10. Let U be a Sylow 2-group of A normalized by Z, and recall $M = M(24)$. We saw in 16.7 that $N_M(U)$ acts on $E = \langle a,b,d\rangle$ with a,b,d in the set D of 3-transpositions and $O^2(C_M(U)) = J \cong U_6(2)/E_4$. By 15.9 , $I \le C_M(S)$ for some D-subgroup $S \cong S_3$. Recall $C_M(S) \cong S_3/P\Omega_8^+(3)$. As S commutes with U, it commutes with E by a property of 3-transposition groups. Then as a,b,d are commuting reflections in $O_8^+(3)$, we conclude $C_M(SU) \cong E \times O_5(3)$. As Z acts on $C_M(SU)$ and centralizes B, we conclude $C_M(SZU) \cong O_5(3)$. On the other hand Z induces an outer automorphism on J centralizing $SC_M(SUZ)$ so $E(C_G(ZU)) \cong \Omega_m(2)$, for m = 6 or 7. Then as $C_G(Z)/O_3(C_G(Z)) \cong \Omega_6^-(3)$ it follows that $E(C_G(ZU)) \cong \Omega_5(3) \cong \Omega_6^-(2)$, so that $S \trianglelefteq N_G(ZU)$. But now $L = \langle A, N_L(ZU)\rangle \le N_M(S)$. This is impossible as $\Omega_8^+(3)$ contains no element of order 21 and hence no subgroup isomorphic to L.

So A moves 5 points of Δ. Hence as $K \cap D$ is the set of transpostions of K, Z is inverted by some $d \in D \cap K$, and by [23], d centralizes $C_G(Z)/Z$; indeed the members of D inverting Z are the members of dZ. In particular d acts on $L = \langle C_L(Z), A\rangle$, so $L\langle d\rangle \cong$

S_n is a D-subgroup. Let A' be an L-conjugate of A in B. By 5.9, $C_G(A)$ is a D-subgroup isomorphic to A_r, where r = 9,7,5 for G = M(24)′,M(23),M(22), respectively. Indeed if G is M(24)′ and n = 11 then r = 9 so working in $C_G(A)$ there is a D-subgroup $S \cong S_3$ centralizing I. Then as $S \leq C_M(A')$ centralizes A, working in $C_G(A')$, an element of $S \cap D$ centralizes $C_L(A')$ and hence also $L = \langle B, C_L(A')\rangle$. On the otherhand if G is M(23) then r = 7 so by 15.10 there is a unique S_{12}-subgroup J of G with $N_G(A) \leq J$. Then as A' is also fused to A in J, if n = 11 we have $L = \langle N_L(A), N_L(A')\rangle \leq J$. Hence as L is determined up to conjugacy in J, (2) holds.

So we may take n = 10. Let $J = C_{L\langle d\rangle}(Z)$, so that $J \cong S_7$ and $B \leq J$. J is a D-subgroup of $C_G(Z) \cong S_3/P\Omega_8^+(3)$, $\Omega_7(3)$, or $Z_2/P\Omega_6^-(3)$, for G = M(24), M(23), or M(22), respectively. As J is a D-subgroup it is generated by reflections on the orthogonal space corresponding to $C_G(Z)$ and hence its commutator space on that module is nondegenerate of dimension 6 and sign -1, and is the nontrivial irreducible in the permutation module for J. In particular if G is not M(22) then J centralizes an element e of $C_D(Z)$. Then working in $C_G(B)$, as e centralizes Z and $C_J(B) \in N_{L\cap D}(A)$, e centralizes A. But then e centralizes $L = \langle J,A\rangle$, completing the proof of (1).

This leaves G = M(22). Here, subject to the identification of D with one of the two classes of reflections in $O_6^-(3)$, there are exactly two $C_G(Z)$-classes of D-subgroups of $C_G(Z)$ isomorphic to J. (cf. 2.4 and 4.10 in [2].) From the structure of $O_6^-(3)$, the normalizer in $C_G(Z)$ of $Q = P\cap J$ acts on J. It follows there are just two such subgroups J and J' containing Q. So as $L = \langle J,A\rangle$ and I is determined up to conjugacy in G, there are at most two classes of A_{10}-subgroups of G. As $N_G(L) \cong S_{10}$ while no involutory outer automorphism of G centralizes an A_{10}-subgroup of G it follows from a Frattini argument that there are exactly two classes fused in Aut(G). Hence the proof is at last complete.

(16.9) (1) Every $\Omega_8^+(2)$-subgroup of G = M(24) or M(23) is contained in a $P\Omega_8^+(3)$-subgroup of G.

(2) There are two G classes of $\Omega_8^+(2)$-subgroups of G with representatives L and K.

(3) $N_G(L) \cong S_4 \times \mathrm{Aut}(L)$ or $Z_2 \times \mathrm{Aut}(L)$ and $N_G(K) \cong S_3 \times O_8^+(2)$ or $O_8^+(2)$, for G = M(24) or M(23), respectively.

(4) $N_G(L)$ and $N_G(K)$ are D-subgroups of G.

(5) Each subgroup of M(24) isomorphic to $Sp_8(2)$ or $Sp_4(4)$ centralizes a 3-transposition of M(24).

Proof. Let $M = M(24)$ and $\Omega_8^+(2) \cong L \leq M$. Let $P \in Syl_5(L)$. If X is a subgroup of P of order 5 which has fixed points on the natural module for L, then $E_X = E(C_L(X)) \cong A_5$ and PE_X is contained in a unique subgroup $L_X \cong A_5 \times A_5$ of L. Further the second component F_X of L_X is fused to E_X in L. As Aut(L) is transitive on subgroups of P of order 5, the same holds for any $1 \neq X < P$. Let $\Delta = (L_X : 1 \neq X < P)$. By 15.9 either E_X is of root type (ie. E_X is invariant under a 3-transposition and $C_M(E_X) \cong S_9$) or $N_M(E_X) \leq I_X \cong S_3 \times S_6 \times S_6$ with $C_M(E_X) \cong S_3 \times S_6$. In particular if L_X is of root type then $S_4 \cong S = O^5(C_M(P)) = C_M(L_X)$ while otherwise $S_3 \cong S_X = C_G(I_X^\infty) = C_M(L_X)$. But $L = \langle A,B\rangle$ for distinct A,B in Δ, so if any member of Δ is of root type then L centralizes an S_3-subgroup of S. Even if none is of root type each pair A,B from Δ centralizes a 3-transposition d_{AB} in $S_A \cap S_B$ (Since $S \cong S_4$ and $S_3 \cong S_A \leq S$, we see $S_A \cap S_B \cap D$ is nonempty.) and then either $S_3 \cong S_A = \langle d_{AB}, d_{AC}\rangle$ centralizes L or $d_{AB} = d_{AC}$ in which case $S_C = S_A$ or S_B centralizes L. So in any case L is centralized by an S_3-subgroup T of S.

Finally $N_M(T) = N_M(J)$ where $J = E(N_M(T)) \cong P\Omega_8^+(3)$ and from remarks above it is clear that $C_M(L) = T$ or S. However $C_M(S) \cong Aut(\Omega_8^+(2))$ while $S_3/P\Omega_8^+(3) \cong N_M(T)/T$ has two classes of $\Omega_8^+(2)$-subgroups.(cf. 9.8 or [15]) Hence the first four parts of the lemma follow since M(23) is the centralizer of a 3-transposition of M.

A similar argument establishes part (5). Here let I be a subgroup of M isomorphic to $Sp_8(2)$ or $Sp_4(4)$ and containing P. In the first case I has subgroups U and V containing P and isomorphic to $\Omega_8^+(2)$ and $A_6 \times A_6$, and in the second isomorphic to $A_5 \times A_5$. Indeed $L = \langle U,V\rangle$ and arguing as above U and V centralize a common 3-transposition of M.

(16.10) M(24) has no M(22)-subgroup and one class of M(23)-subgroups.

Proof. Let $M = M(24)$ and $L \leq M$ with L isomorphic to M(23) or M(22). Then there is a subgroup K of L isomorphic to S_n, $n = 12$ or 10, and by 16.8 and 15.9, K is contained in a D-subgroup X of M isomorphic to $Z_2 \times S_{12}$. It follows that the 3-transpositions of K, and hence also those of L, are in D. Hence if Z is of order 3 in K with $C_K(Z) \cong Z_3 \times S_{n-3}$ then $C_M(Z) \cong Z_3 \times (S_3/P\Omega_8^+(3))$ and $C_L(Z) \cong Z_3 \times \Omega_7(3)$ or $Z_3 \times (Z_2/U_4(3))$, respectively, with $E(C_L(Z))$ of index at most 2 in a D-subgroup of $C_M(Z)$. As $O_8^+(3)$ contains no subgroup generated by reflections and isomorphic to $U_4(3)$ (as opppsed to $\Omega_6^-(3) \cong U_4(3)/Z_3$) it follows that M has no M(22) subgroup. Thus L is M(23). Then $E(C_K(Z)) \cong A_9$ so as the 3-transposition d in Z(X) is in $C_M(C_K(Z))$, it centralizes the unique D-subgroup Y of $C_M(Z)$ containing $E(C_K(Z))$ with E(Y) isomorphic to $\Omega_7(3)$. That is d centralizes $E(C_L(Z))$ and hence also $L = \langle K,E(C_L(Z))\rangle$. Hence $L = C_M(d)$ is determined up to conjugacy in M.

(16.11) Every ${}^2F_4(2)'$ subgroup of M(23) and M(24) centralizes a 3-transposition in the respective group.

Proof. Let M be M(24) or M(23) and ${}^2F_4(2)' \cong L \le M$. Let $P \in Syl_5(L)$; then $N_L(P)/P \cong Z_4{*}SL_2(3)$. Let z be an involution in $N_L(P)$ with $1 \ne X = C_P(z)$. Then z is 2-central in L. Let $I = O^2(C_M(z)$, $Y = O^2(C_L(z))$, and $J = O_2(Y)$.

Claim first that z is the product of 3 or 4 commuting 3-transpoitions of M, where M is M(23) or M(24), respectively. If M is M(24) then as $z \in [M,M]$, z is the product of an even number of transpositions. Certainly $z \notin D$ as L is not generated by 3-transpositions. So if the claim is false then z is the product of two commuting 3-transpositions. (cf. 13.5 in [13].) But then z is not a square in I because $I \cong M(22)/Z_2$ or $U_6(2)/E_4$ is quasisimple and all involutions in I/Z(I) lift to involutions in I. On the other hand z is a square in Y. (cf. section 3 in [40] where it is also shown J is of class 3 with second center $E \cong E_{32}$.)

Next by [23], $S = C_M(X) \cong S_9$ or S_7 so as z inverts [P,z] and is the product of m commuting transpositions in S, P is centralized by 2 or 1 of these transpostions, say a,b or a, respectively. It remains to show a centralizes Y as then $L = \langle P,Y\rangle \le C_M(a)$.

If M is M(23) this is clear as $O^{5'}(C_M(z)) \le C_M(a)$. So let M be M(24), define J and E as above, and let $Q = O_2(C_G(z))$. Then $Q \cong Q_8^6$ and $H^* = C_M(z)/Q \cong E_4/U_4(3)/Z_3$ acts as a subgroup of $Z_2/SU_6(2)$ in its action on $\tilde{Q} = Q/\langle z\rangle$. As Q is of class 2 and J of class 3, $J \not\le Q$, which forces $J^* \cong E_{16}$ and $N_{H^*}(J^*) \cong S_6/E_{32}$, from $X^* \le N(J^*)$ and the 2-local structure of H^*. Indeed as J centralizes $\tilde{E}$ and a^* centralizes X^* it follows that $\tilde{E}$ is a 2-dimensional totally singular subspace of the 6-dimensional unitary space $\tilde{Q}$ over GF(4) and $N_{H^*}(J^*) = C(a^*)\cap N(J^*) \cong S_5/E_{32}$. Thus I acts on a^Q of order 4 so again $I \le C(a)$ as desired.

(16.12) Let G be M(22) and A = Aut(G). Then

(1) P is contained in exactly two conjugate subgroups M and M^b of G isomorphic to $^2F_4(2)'$. $N_A(M) \cong {}^2F_4(2)$.

(2) $N_G(P)$ is transitive on the set of subgroups of G containing P and isomorphic to $L_2(25)$. M contains such a subgroup L, L is of index 2 in $N_A(L)$, and $N_A(L)$ has semidihedral Sylow 2-groups.

Proof. First by 15.17, A contains a subgroup isomorphic to $^2F_4(2)$.

I claim next that if X is a subgroup of G isomorphic to the Tits group or $L_2(25)$ then X is selfnormalizing in G. This follows from a Frattini argument and the fact that if $Y \in Syl_{13}(X)$ then $|N_X(Y)|_2 = 2 = |N_G(Y)|_2$.

Next there is a subgroup $V \cong Z_4$ of G such that $N_G(P) = BP$ where $B = N_G(P) \cap C(V)$, $B/V \cong S_4$, and $O^2(B)V \cong V*SL_2(3)$. We have seen that G contains a selfnormalizing subgroup M isomorphic to the Tits group. From Conway-Wales [20], M contains a subgroup K isomorphic to $L_2(25)$. $N_M(P)/P \cong Z_4*SL_2(3)$ so $N_M(P) = VO^2(B)P$. Now it is easy to see (cf. 16.16) that $M = \langle N_M(P),K\rangle$. Thus (2) implies (1), so it remains to show (2).

To begin observe BP is transitive on its cyclic subgroups of order 12 and that if U is such a subgroup then U contains a P-conjugate of V, which we take to be V. Let X be of order 3 in U. We saw in the preceeding paragraph that $X \le M$. Now M has one class of elements of order 3 and $N_M(X) \cong D_8/3^{1+2}$. In particular if $R \in Syl_3(C_M(X))$ then $X \le \Phi(R)$. This last fact says X is not in 2 of the 4 classes of subgroups of G of order 3, so either $|C_G(X)| = 8.3^7$ or X is 3-central in G. I claim the former is true, so assume the latter. We will see in section 19 that $Q = O_3(C_G(X)) \cong 3^{1+6}$ and $G \le E \cong {}^2E_6(2)$ with $O_{3,2}(C_E(X)) = H_1*H_2*H_3$, where $H_i \cong SU_3(2)$ and $N_E(X)$ permutes the groups H_i, $1 \le i \le 3$. Indeed $O^3(C_G(X))$ is the subgroup of $N_G(X)$ fixing each of the groups $Q_i = O_3(H_i)$, $1 \le i \le 3$, and $C_A(X)$ is $C_G(X)$ extended by an involution permuting the Q_i's nontrivially. Thus if a is an involution in $C_A(X)-G$ then $Q_i^a \ne Q_i$ for some i so $Q_iQ_i^a \cap C(a) \cong 3^{1+2}$. In particular if $S \in Syl_3(C_G(X\langle a\rangle))$ then $X \le \Phi(S)$.

By 15.10 there is a subgroup N of G isomorphic to $Aut(\Omega_8^+(2))$ containing P and indeed the proof shows $N = C_G(a)$ for some involution a in A-G. In particular as a centralizes X, $X \le \Phi(S)$. Hence as $N/E(N) \cong S_3$, $X \le E(N)$. This is impossible as $N_{E(N)}(P)$ is a $3'$-group.

So the claim is established. In particular from the list of conjugacy classes of

G, U is determined up to conjugacy in G by its order and the fact that $X \le U$; moreover $|C_G(U)| = 36$. Thus $C_G(U) = V \times Y$ where Y is of order 9.

Suppose now that $L_2(25) \cong L \le G$; conjugating in G if necessary we may take $PU \le L$. Now there is an involution t in L inverting U and $L = \langle P,W \rangle$, where $W = U\langle t \rangle$ is dihedral of order 24. Also there is $b \in B$ inverting X. Then $N_B(X) \le N_B(U)$, so $t,b \in N(U) \le N(Y)$ and $VY\langle t \rangle/U \cong Z_6$ or S_3, with W the unique D_{24}-subgroup containing U in the first case and with three such subgroups W_i, $1 \le i \le 3$, in the second. As L is selfnormalizing in G, $L \ne L^b$ so $W \ne W^b$. Thus the second case holds and with suitable notation, $W = W_2$, $W_3 = W^b$, and $\langle P,W_1 \rangle \not\cong L_2(25)$ as b acts on W_1. This completes the proof of (2) and hence of the lemma.

(16.13) Let G be Sz and $A = Aut(G)$. Then

(1) G has two classes of $U_3(4)$-subgroups fused in A. If K is such a subgroup then $N_G(K) \cong Aut(K)$ and $N_G(K)$ is contained in a $G_2(4)$-subgroup of G.

(2) Up to conjugation G has a unique $G_2(4)$-subgroup L. L is selfnormalizing in G and has two classes of $U_3(4)$-subgroups fused in $N_A(L) \cong Aut(L)$.

(3) Up to conjugation G has a unique J_2-subgroup M. $N_G(M) \cong Aut(M)$. $M \le L$ but $N_G(M) \not\le L$.

Proof. We saw in 16.5 that G has a $G_2(4)$-subgroup while from the Lie theory $G_2(4)$ has a $U_3(4)$-subgroup.

So let $U_3(4) \cong K \le G$, $T \in Syl_2(K)$, and X of order 5 in $N_K(T)$. Then $T = [T,X]$ and X centralizes $Z = Z(T) = \Omega_1(T) \cong E_4$. Let $z \in Z^{\#}$. As $z \in O^2(C_G(z))$, z is 2-central in G. (cf. [23]) As X centralizes Z, $Z \le O_2(C_G(z))$, and then Z is determined up to conjugacy in G with $H/R \cong S_3 \times L_2(4)$, where $H = N_G(Z)$ and $R = O_2(H)$ is special of order 4^5 with center Z. (cf section 30) Let $\bar{R} = R/Z$. $End_{GF(2)X}(\bar{R}_1) = GF(16)$, for $\bar{R}_1$ an irreducible GF(2)X-submodule of $\bar{R}$, so the S_3 factor of H/R is in $O^{2'}(C_{GL(\bar{R})}(X)) = L_2(16)$ and hence the subgroup of order 3 in $C_H(X)$ fixes just 2 of the 17 GF(2)X irreducibles on $\bar{R}$ and these 2 are permuted transitively by the S_3-factor. On the other hand $\bar{T}$ is one of these factors since $N_K(T)/T$ is of order 15, so given ZX there are just two choices for T conjugate in $N_G(XZ)$.

Next $N_K(X) \cong Z_5 \times L_2(4)$ and as X centralizes a 2-central involution, $C_G(X) \cong X \times A_6$.Hence Z is contained in exactly two $L_2(4)$-subgroups K_1 and K_2 of $C_G(X)$. Further there is a of order 4 in $N_A(X)$ centralizing $E(C_G(X))$. As a induces an automorphism of order 4 on X centralizing an element of order 3 in $N_G(ZX)$, a induces a field automorphism on $O^{2'}(C_{GL(\bar{R})}(X))$. Thus a interchanges the two conjugates of T under $N_G(XZ)$, while a^2 acts on each conjugate. It follows that K = $\langle K_1,T\rangle$ is determined up to conjugation in A. Also $K\langle a^2\rangle = N_G(K) \cong Aut(K)$.

Similarly if $G_2(4) \cong L \leq G$ then, conjugating in G if necessary, $I = O^2(N_G(Z))$ is a parabolic of L and $L = \langle I,C_L(X)\rangle$ with $C_L(X) \cong Z_5 \times L_2(4)$ determined up to conjugacy in $N_G(I)$. Thus L is determined up to conjugacy in G and as no involutory automorphism of G centralizes L, a Frattini argument shows $L = N_G(L)$ while $N_A(L) \cong Aut(L)$. In particular we may take $K \leq L$. As a acts on I and K_1, a acts on $L = \langle I,K_1\rangle$. Now K^a is fused to K in $N_A(L)$ but not G. So (1) and (2) are established.

Finally there is an involutory outer automorphism of G whose centralizer is $Aut(J_2)$. On the other hand suppose $J_2 \cong M \leq G$ with $P \leq M$. The set Δ of subgroups X of P of order 5 with $E(C_M(X) \cong A_5$ is of order 3. If Z is of order 3 in $C_M(X)$ then Z is 3-central in M so $Z \leq C_M(Z)^{\infty} \cong A_6/Z_3$. This forces $C_G(Z) \cong U_4(3)/Z_3$ and this in turn implies $E(C_G(X)) \cong A_5$. Thus $E(C_G(X)) = E(C_M(X))$. Proceeding in an inductive context, we know the overgroups of P in M and hence conclude $M = \langle E(C_M(X)) : X \in \Delta\rangle$. Thus the first the first two statements of (3) are established. It is known that $G_2(4)$ contains a J_2-subgroup, but as $|N_L(P)|_2 = 4$ while $|N_G(M)\cap N(P)|_2 = 8$, $G_2(4)$ does not contain $N_G(M)$.

(16.14) Let G = HS, X of order 3 in P, and α an involutory outer automorphism of G centralizing $N_G(X)$. Then

(1) $N_G(X) \cong S_3 \times S_5$. Let $Y = P\cap E(N_G(X))$ and W a third subgroup of P of order 3.

(2) There are exactly two subgroups of $C_G(X)$ isomorphic to $Z_3 \times A_4$ containing P; these subgroups are fused in $C_G(X)$. Let A be one of these groups and A_i and B_i, i = 1,2, be conjugates of A under N(P) in $C_G(Y)$ and $C_G(X)$, respectively.

(3) $L_1 = \langle A,A_1\rangle \cong A_4 \times A_4$, $L_2 = \langle A,A_2\rangle \cong A_7$, and $L = \langle L_1,L_2\rangle \cong A_8$.

(4) $L = E(C_G(\alpha))$ and $N_G(L) \cong S_8$.

(5) $\langle A,B_i\rangle = K_i \cong A_7$ and $B_1^{\alpha} = B_2$.

(6) $\langle L_1,K_1\rangle = I \cong M_{22}$ and $\langle L_2,K_i\rangle = J_i \cong U_3(5)$. I is selfnormalizing while $N_G(J_i) \cong Z_2/U_3(5)$.

(7) The representatives for the conjugacy classes of G isomorphic to A_7, A_8, M_{22}, and $U_3(5)$ are L_2, K_i, L, I, and J_i. The obvious fusion takes place in Aut(G) via α.

Proof. The first remark comes from [23] as does the observation that $C_G(\alpha) \cong S_8$. Part (2) is immediate from part (1) while parts (3) and (4) follow from $C_G(\alpha) \cong S_8$.

Let D be an A_7-subgroup of G; conjugating in G if necessary we may assume $A \le D$. Hence $D = \langle A,B\rangle$, where B is A_i or B_i, or $D = \langle A,B_i^t\rangle$ for $t \in (N_G(A)\cap N(P)) - N(W)$. So as $D \ne L_1$, there are at most 3 classes of A_7-subgroups of G and in case of equality L_2 and K_i, $i = 1,2$, are representatives for these classes.

It is well known that G contains an M_{22}-subgroup I and a $U_3(5)$-subgroup J with $N_G(J) \cong Z_2/U_3(5)$. Conjugating in G if necessary we may take $A \le I$ and J. As an involutory outer automorphism of G centralizes no $U_3(5)$-subgroup it follows from a Frattini argument that J is not conjugate to J^α in G. Next J has 3 classes of A_7-subgroups and $Out(J) \cong S_3$ is transitive on these classes, while J has one class of elements of order 3, so we conclude from a Frattini argument that A is contained in 3 A_7-subgroups of J. From the last paragraph one of these is L_2 and we may take the others to be K_1 and K_1^t. Thus $J = J_1$. It is easy to see $B_1^\alpha = B_2$, so (5) holds. Also as $J = \langle L_2,K_1\rangle$ and I is generated by L_1 and either of two A_7-subgroups containing A it follows that this subgroup is B_i for some i. Further at this point the remaining parts of the lemma are reasonably clear.

(16.15) Up to conjugacy $G = M(23)$ contains a unique $Sp_4(4)$ subgroup L. $N_G(L) \cong Aut(L)$.

Proof. Let $Sp_4(4) \cong L \le G$; we may take $P \le L$. Then there are two subgroups I and J of L isomorphic to $A_5 \times A_5$ and containing P. One is generated by long root groups of L and the other by short root groups, so they are fused in Aut(L) but not in L. Let A be a component of I and $U \in Syl_2(A)$. As usual either U is normal in a D-subgroup $S \cong S_4$ of G or $U \le E \cong E_8$ with E generated by three 3-transpositions a,b,d. $N_L(U) = BYV$ where Y is of order 3 in $N_A(U)$, B is the second component of I, $V = O_2(BYV) = [B,V] \cong E_{64}$, and $L = \langle A,V\rangle$. As we've seen before, (cf. 16.9.3) if $U \trianglelefteq S$ then $N_{M(24)}(U) \cong S_4 \times Aut(\Omega_8^+(2))$ so $U\cap N_G(U)^\infty = 1$. So $U \le E$. In particular by 15.10.2, I is contained in a D-subgroup $I_0 \cong S_{12}$ of G and if Δ is a set of 12 points permuted by I_0, then the subsets Δ_A and Δ_B of Δ moved by A and B are of order 6 with $\Delta = \Delta_A \cup \Delta_B$. In particular I is determined up to conjugacy in I_0, and hence determined up to conjugacy in G by 15.10.1. Similarly J is conjugate to I in G. Also

$H = N_G(U) = C_G(t)$ where $t = abd$. So $H \cong Z_2/GU_4(2)/(E_4 \times Q_8^4)$. Let $Q = O_2(H)$, $H^* = H/Q$, and $\tilde{H} = H/E$; then $\tilde{Q}$ is the natural GF(4)-module for $O^2(H^*) \cong GU_4(2)$.

As V/U is the natural module for $B = L_2(4)$, it follows that $\tilde{V}$ is a totally singular plane of the unitary space $\tilde{Q}$ and that there is just one other such B-invariant subspace $\tilde{V}^b$, where $b \in C(A)$ induces an outer automorphism on B. As V is the unique B-invariant complement to $\langle t\rangle$ in EV, BV is determined up to conjugacy in $N_G(I) \cap C(U)$, so $L = \langle V,A\rangle$ is determined up to conjugacy in G. Indeed this shows $N_G(I)$ is transitive on the G-conjugates of L containing I, so $N_G(L)$ is transitive on the G-conjugates of I contained in L. In particular as J is such a conjugate it follows that $N_G(L) \cong Aut(L)$.

(16.16) Let $L = O^p(L)$ be a proper subgroup of G containing P such that $F^*(L)$ is a nonabelian simple group. Then L is simple and one of the following holds:

(1) G is $M(24)'$ and $L \cong A_n$, $10 \le n \le 12$, M(23), He, $Sp_8(2)$, $Sp_4(4)$, $\Omega_{10}^+(2)$, $\Omega_8^+(2)$, $\Omega_8^+(3)$, ${}^2F_4(2)'$, or $L_2(25)$.

(2) G is M(23) and $L \cong A_n$, $10 \le n \le 12$, $Sp_8(2)$, $Sp_4(4)$, $\Omega_8^+(3)$, $\Omega_8^+(2)$, ${}^2F_4(2)'$, or $L_2(25)$.

(3) G is M(22) and $L \cong A_{10}$, $\Omega_8^+(2)$, ${}^2F_4(2)'$, or $L_2(25)$.

(4) G is HS and $L \cong A_n$, $6 \le n \le 8$, M_{11}, M_{22}, $U_3(5)$, or $L_3(4)$.

(5) G is Sz and $L \cong G_2(4)$, J_2, $U_3(4)$, or $L_2(25)$.

(6) G is F_2 and $L \cong F_3$, ${}^3D_4(2)$, $F_4(2)$, or $L_2(49)$.

(7) G is F_3 and $L \cong {}^3D_4(2)$.

(8) G is He and $L \cong Sp_4(4)$.

Proof. Observe first that as P is abelian and $L = O^p(L)$ with $F^*(L)$ nonabelian simple, L is indeed simple; for from the structure of $Aut(L_0)$, where L_0 is simple, if $L_0 \le L_1 \le Aut(L_0)$ and $E_{p^2} \cong P \in Syl_p(L_1)$ then $P \le L_0$.

Suppose $L \cong A_n$. If G is ON or J_1 then P is abelian of order 81 or 8, whereas no alternating group has such a Sylow group. So these cases are out. In the remaining cases $P \cong E_p2$ so $2p \le n < 3p$ and G has an $A_m \times A_{n-m}$ subgroup for $m < n$. We conclude from these observations together with 15.3, 15.4, 15.9, 15.10, 16.4, and 16.5 that the lemma holds or G is He or Sz. As the order of A_{10} does not divide that of He we may assume G is Sz and $n = 10$. But now for $A \times B \le L$ with $A \cong B \cong A_5$, $A \in B^L$, whereas by 16.5, A cannot be fused to B in G.

Suppose next that L is sporadic. Then L is one of the groups considered in this section. As the order of L divides that of G it follows that either the lemma holds or one of the following holds: (i) G is F_2 and L is Co_1. (ii) L is J_2 and G is one of the groups of 16.1.3. (iii) G is a Fischer group and L is Sz. Observe 16.10 is used to eliminate the case $L \cong M(22)$, $G \cong M(24)'$ or $M(23)$. As A_7 is not a subgroup of $L_3(4)$, Co_1 is not a subgroup of F_2 by 16.2. Case (iii) is out by 16.7 and 16.13, since each of the Fischer groups is a section of M(24).

This leaves $L \cong J_2$. Suppose first that G is He, and let z be a 2-central involution in L and $H = C_L(z)$. Then $H \cong A_5/Q_8{*}D_8$. In particular z is a square in $C_G(z)^{\infty}$ so z is 2-central in G, impossible as 5 does not divide the order of the centralizer of a 2-central involution of He. Hence we may assume G is a Fischer group, and indeed take $G = M(24)'$ as each Fischer group is a section of $M(24)'$.

Let Z be a 3-central subgroup of L of order 3; hence $Z \leq C_L(Z)^{\infty}$, so Z is even 3-central in G. On the other hand if X is of order 5 in L with $K = E(C_L(X)) \cong A_5$ then subgroups of K of order 3 are 3-central in L and hence also in G. But $K \leq E(C_G(X)) \cong A_9$ and hence moves 5 or 6 of the 9 points permuted by that group. In either case we have seen earlier in this section that a subgroup of K of order 3 is not 3-central in G.

Assume now that $L \in Chev(p)$. Then as P is Sylow in L, $L \cong L_2(q)$, where $q = |P|$. In particular L, and hence also G, has an element of order $(q+1)/2$, (or 9 if G is J_1) which reduces us to the cases listed in the lemma. (Sylow 5-groups of Co_1 and F_3 are of exponent 5.)

This leaves us in the situation where $L \in Chev(r)$ for some prime $r \neq p$. If G is J_1 then as $P \cong E_8$, L is a Ree group; but then the order of L does not divide that of G. So p is odd. Let $R \in Syl_r(L)$. As P is noncyclic $|R| \geq r^3$, and in case of equality $L \cong L_2(r^3)$ with $p = 3$, or $L_3^{\epsilon}(r)$ with $r \equiv \epsilon \bmod p$. It follows that either $r \leq 3$ or G = HS and $L \cong U_3(5)$, and we may assume the former.

If p = 3 then G is HS or ON and r = 2. As L has abelian Sylow 3-groups of order 9 or 81 and |L| divides |G|, it follows that G is HS and $L \cong L_3(4)$ or $L_4(2)$, so that the lemma holds.

Therefore $p \geq 5$. Hence by Seitz [34], either $L = \langle O^{r'}(C_L(x)) : x \in P^{\#}\rangle$ or p = 5 and L is the Tits group. Observe that if G is Sz then $Aut_G(P) \cong Z_4 \times S_3$ which contains no $Z_4{*}SL_2(3)$ subgroup. But if L is the Tits group then $Aut_L(P) \cong Z_4{*}SL_2(3)$

so G can't be Sz. Also 2^{11} divides the order of the Tits group but not that of J_2 or He. Thus we can assume L is not the Tits group. So for some $x \in P^{\#}$, $K = O^{r'}(C_L(x)) \neq 1$. By 9.1, $K \in Chev(r)$. As K is a P-invariant subgroup of $O^{r'}(C_G(x))$ it follows from 16.2 that either $r = 2$ and $K \cong L_2(4)$, $Sp_4(2)$, $L_3(2)$, or $L_4(2)$, or $r = 3$, G is Sz, M(23), or M(24)′, and $K \cong L_2(9)$.

Suppose $r = 3$. By 9.1, L is over GF(3) or GF(9). As $P \cong E_{25}$ is Sylow in L, as each component of the centralizer of an element of order 5 is $L_2(9)$, and as $|L|$ divides $|G|$ it follows that $L \cong \Omega_8^+(3)$ and G is M(23) or M(24)′, so the lemma holds.

So $r = 2$ and by 9.1, either L is over GF(2) or $K \cong L_2(4)$ and L is over GF(4). In any event either $p = 5$ and G is He, Sz, J_2, or a Fischer group, or $p = 7$ and G is Co_1, F_3, or F_2.

Suppose $p = 7$. As components of centralizers of elements of order 7 are $L_3(2)$, as E_{49} is a Sylow 7-group of G, and as the order of L divides that of G it follows that either the lemma holds or G is F_2 or F_3 and L is $L_6(2)$. But then L has a proper parabolic containing P whereas by 16.3, $\mathcal{U}_G(P,2) \subseteq C_G(P)$.

Therefore $p = 5$, G is He, Sz, J_2, or a Fischer group, and K is $L_2(4)$ or $Sp_4(2)$. In particular L is defined over GF(2) or each component is $L_2(4)$ and L is defined over GF(4). Proceeding as usual we conclude that either the lemma holds or the pair (G,L) is one of a small number of possiblities. Most of these possibilities are eliminated using 16.7. $\Omega_8^+(2)$ is not contained in Sz as Sz has no E_{27}/Q_8^4 subgroup. These remarks are sufficient to complete the proof of the lemma.

(16.17) If M is a maximal subgroup of G containing P then one of the following holds:

(1) $M = N_G(X)$ for some $1 \neq X < P$.

(2) G is HS, M(23), M(24)′, or F_2 and $M = N_G(X)$ for some $1 \neq X \leq O^p(C_G(P))$.

(3) $F^*(M)$ is a nonabelian simple group.

(4) $F^*(M) \in \Gamma$.

(5) G is HS, $M \cong S_6/E_{16}$, and M is determined up to conjugacy in G.

Proof. We may assume (1) does not hold and hence that $O_p(M) = 1$. If $R = O_{p'}(M) \neq 1$ then by 16.3 either (2) holds or G is HS, $R = O_2(M)$, and $[P,R] \neq 1$. We may assume the latter. Then $R = \langle C_R(X) : X \in \Delta\rangle$, where Δ is the set of subgroups X of P of order 3 with $C_R(X) \neq 1$. But for $X \in \Delta$, $C_G(X) \cong Z_3 \times A_5$ so $C_R(X)P \cong Z_3 \times A_4$. In particular by 16.14, $|\Delta| \leq 2$ and in case of equality $R \cong E_{16}$ is determined up to conjugacy in G. Moreover in that case by 16.14 there is an M_{22}-subgroup I of G containing PR and by [19], $N = N_I(R) \cong A_6/E_{16}$. Let $T \in Syl_2(N)$ and $Z = Z(T)$. It is easy to see that Z is 2-central in G and $C_N(Z)$ is maximal in $C_I(Z)$. From this embedding and the structure of $C_G(Z) = K$ we see $O_2(K)C_N(Z) = N_K(R)$ and $|N_K(R):C_N(Z)| = 2$. Then as N is transitive on $R^{\#}$ it follows that $M = NO_2(K)$ and $|M:N| = 2$. Hence $M \cong S_6/E_{16}$ and as R is determined up to conjugacy, so is M.

Of course if X is the unique member of Δ then R is of order 4 and determined up to conjugacy in G. Indeed it is easy to see $N_G(R) \cong S_4 \times S_4$ is properly contained in $N_G(O_2(R))$, contradicting the maximality of M.

So $O_\infty(M) = 1 = O_{p'}(M)$. Therefore either (3) or (4) holds, completing the proof.

Observe that the results of this section and the previous section essentially establish the validity of the tables in the introduction. Namely 16.17 identifies four categories of potential maximal subgroups. Lemmas 16.4 and 16.5 together with results in section 15 determine the members of Γ. Results in sections 15 and 16 determine the maximal subgroups M with $F^*(M)$ simple, except in certain cases when $F^*(M) \cong L_2(p^2)$ (Or M_{11} when G is HS.) These maximal subgroups appear to be too small to easily investigate via the methods of this section.

17. Groups of GF(p)-type.

In this section p is a prime, G is a group of GF(p)-type, Q is a large symplectic subgroup of G, and $\langle z\rangle = Z = \Omega_1(Z(Q))$. Recall this means:

(L1) Q is a p-subgroup of G of symplectic type and exponent p or 4.

(L2) $C_G(Q) \leq Q$.

(L3) $Q \trianglelefteq N_G(Z)$.

(L4) $\mathcal{N}_G(Q,p') = 1$.

By a result of P. Hall (cf. Chapter 8 in [6]) p-groups Q of symplectic type and exponent p or 4 are known. In particular Q is the central product of Z(Q) with w extraspecial subgroups of order p^3. The parameter w is called the *width* of Q. If p is odd then Q is extraspecial and its isomorphism type is determined by w. Write p^{1+2w} to denote the unique extraspecial p-group of exponent p and width w. If p = 2 then up to isomorphism there are two extraspecial groups of width w denoted by D^w and $D^{w-1}Q$. The notation D^nQ^m denotes the central product of n copies of the dihedral group of order 8 with m copies of the quaternion group of order 8 and with all centers identified. The crucial fact here is that $D^2 \cong Q^2$. Write 2^{1+2w} to denote either of the extraspecial 2-groups of with w and exponent 4. In addition there is a unique 2-group of symplectic type, width w and exponent 4 which is not extraspecial. This is the group $Z_4{*}D^w \cong Z_4{*}D^{w-1}Q$.

(17.1) Let S be a p-subgroup of G containing Q. Then

(1) $Z(S) \leq Z(Q)$, and

(2) Q is weakly closed in S with respect to G.

Proof. Observe L2 implies (1) while 2.3 implies (2).

(17.2) Assume P is a p-subgroup of a group H satisfying L1, L2, and L3. Assume further there exists x of order p in P - Z(P) with $\mathcal{N}_{C_H(x)}(C_P(x),p') \subseteq C_H(\Omega_1(Z(P)))$.

Then P also satisfies L4.

Proof. Let $\langle z\rangle = \Omega_1(Z(P))$, $X = \langle x,z\rangle$, and $1 \neq R \in \mathcal{N}_G(P,p')$. Then $C_R(y) \neq 1$ for some $y \in X^\#$. So as P is transitive on the subgroups of order p in X distict from $\langle z\rangle$, we may take y = x or z. In either event $C_R(y) \leq C(z)$. Hence by L3, $[C_R(y),P] \leq R\cap P = 1$, contradicting L2.

(17.3) Assume P is a p-subgroup of a group H satisying L1, L2, and L3. Assume further there exists x of order p in P - Z(P) with $z \in O_p(C_H(x))$. Then P also satisfies L4.

Proof. For $R \in \mathcal{N}_{C_H(x)}(C_P(x),p')$ we have $[R,z] \leq R\cap O_p(C(x)) = 1$, so 17.2 completes the proof.

(17.4) Let $L = \langle z^G\rangle$. Then one of the following holds:

(1) $Q \trianglelefteq G$.

(2) L is abelian and $F^*(G) = O_p(G)$.

(3) $L = F^*(G)$ is a nonabelian simple group.

(4) p = 2, $Q \cong D^2$, and $F^*(G) \cong A_5 \times A_5$.

Proof. This is essentially immediate from 8.1.

(17.5) Assume $g \in G$ with $z^g \in Q - Z$. Let $X = \langle Q,Q^g\rangle$ and $V = \langle z,z^g\rangle$. Then $z \in Q^g$, $V \cong E_{p^2}$, $V \trianglelefteq X$, and $X/C_X(V) \cong SL_2(p)$.

Proof. Q is transitive on the subgroups in V of order p distinct from Z, so if $N_G(V) \not\leq N_G(Z)$ then $O^{p'}(Aut_G(V)) = SL_2(p)$, and the lemma holds. Thus we may assume $N_G(V) \leq N_G(Z)$. In particular this means $z \notin Q^g = P$, since otherwise by symmetry $N_G(V) \leq N(Z^g)$, which is not the case. Let $H = N_G(P)$, $t = z^g$, and $H^* = H/Z(P)$. As $N_G(V) \leq N_G(Z)$, $C_{P^*}(z) = C_P(z)^*$.

Assume the width w of Q is not 1. Then $C_Q(t) = R \cong Z_p \times (Z(Q)*p^{1+2u})$, where u = w-1. $[R\cap P,R] \leq Z\cap P = 1$, so as $V = \Omega_1(Z(R))$, $R\cap P = \langle t\rangle$. Thus $[C_P(z),R] \leq P\cap R = \langle t\rangle$, so $[C_P(z),R^*] = 1$. Let $Y^* = C_{P^*}(z)R^*$. Then $R^* \cong Z(Q)*p^{1+2u}$ and $Y^* = C_{P^*}(z) \times R^*$ so Z^*

$= \Phi(Y^*)$. Hence $N_{P^*}(Y^*) \leq C_{P^*}(z^*) \leq Y^*$, so $P^* = C_{P^*}(z^*) = C_P(z)^*$, contradicting L2.

The same argument works if $Z(Q)$ is not of exponent p, so we can take $Q \cong p^{1+2}$. As $C_{P^*}(z) = C_P(z)^*$, $U = C_P(z)V \cong E_{p^3}$. Indeed $T = VP \in Syl_p(H) \subseteq Syl_p(G)$ and $U = J(T)$. Therefore U is weakly closed in T so $M = N_G(U)$ controls fusion in U. Notice $X \leq M$ as $U = J(T')$ for $U \leq T' \in Syl_p(N_G(Z))$. Also $\langle t\rangle$ and $P \cap U$ are the only nontrivial proper P-invariant subspaces of U. Finally notice Z and $Q \cap U$ are the only proper nontrivial Q-invariant subspaces of U. Therefore X, and hence also M, is irreducible on U. Now as $T/U \in Syl_p(M/U)$ we conclude that p is odd and then conclude that $O^{p'}(Aut_G(U)) \cong \Omega_3(p)$ acts naturally on U. But now Z is weakly closed in $Q \cap U$ with respect to M, (ie. Z is the unique singular point in the subspace $Q \cap U$ of the orthogonal space U.) impossible as $t \in Q \cap U = V$ and M controls fusion in U.

If $z^g \in Q - Z$ write $Z+Z^g$ for the set of subgroups of order p in $\langle z,z^g\rangle$. $Z+Z^g$ is the *line* through the *points* Z and Z^g. If $Q \leq H \leq G$ let $\mathcal{S}(H) = (\mathcal{P}(H),\mathcal{L}(H),\subseteq)$ be the partial linear space with point set Z^H and line set

$$\mathcal{L}(H) = \{Z^g+Z^h : g,h \in H,\ z^h \in Q^g - Z^g\}$$

and with incidence defined by inclusion. Evidently H induces a group of automorphisms of $\mathcal{S}(H)$.

(17.6) Let $Q \leq M \leq G$, $V = \langle z^M\rangle$, $X = \langle Q^M\rangle$, $\Delta \subseteq Q^M$ with $X = \langle\Delta\rangle$, and $R = \prod_{P\in\Delta} C_P(V)$. Assume V is abelian. Then

(1) $C_X(R)R \leq O_p(X) \leq C_X(V)$, $[X,R] \leq R$, and $Z^M = Z^X$.

(2) X acts irreducibly on V.

(3) If $V \leq Q$ then $\mathcal{S}(M)$ is projective space, $X/C_X(V) = SL(V)$, and $Q/C_Q(V)$ is the full group of transvections of V with center Z. Further either $R = C_X(V) = O_p(X)$ or $p = 2$ and $X/R \cong L_4(2)/Z_2$.

(4) If $|V:V\cap Q| = p$ and $\mathcal{L}(M)$ is nonempty then $\mathcal{S}(M)$ is a Shult space. If the rank of this space is at least 3 then V is an orthogonal space (or $p = 2$ and V is a symplectic space) with singular points $\mathcal{P}(M)$, $Q\cap V = Z^{\perp}$, $X/C_X(V) = \Omega(V)$ (or $Sp(V)$), and $Q/C_Q(V)$ is the radical of the parabolic fixing Z. Further either $R = C_X(V) = O_p(X)$ or $X/R \cong \Omega_8^+(2)/Z_2$ or $\Omega_7(3)/Z_3$.

(5) If $[Q,V] = Z$ then $V \leq Q$.

(6) Assume either

(a) $m(Q\cap V) = 2$, or

(b) $|Q:C_Q(V)| = p$.

Then (b) implies (a) and either

(i) V is generated by some member of $\mathcal{X}(M)$, or

(ii) p is odd, $V \cong E_{p^3}$ is the natural module for $X/C_X(V) \cong \Omega_3(p)$, $\mathcal{X}(M)$ is empty, and $|C_X(V):R| \leq 2$.

Proof. As Q is weakly closed in a Sylow p-group of M, $Z^X = Z^M$ by a Frattini argument. Also $Z \leq V$ so $R \leq X \leq C_X(Z) \leq N_G(Q)$. In particular R is a p-group. Further $[C_X(V),Q] \leq C_Q(V)$ so $[X,R] \leq R$. So as $Z^X = Z^M$, $V \leq R$. If Y is a p'-subgroup of $C_X(R)$ then $[Q,Y] \leq C_Q(V) \leq C(Y)$, so $Y \leq C_G(Q) = Z(Q)$. Thus $C_X(R)R \leq O_p(X)$ and hence (1) is established. As $Z = C_V(Q)$ and $V = \langle Z^X\rangle$, (2) holds.

Assume $V \leq Q$. Then $Q/C_Q(V)$ is the full group of transvections with center Z so by 6.6, $X/C_X(V) = SL(V)$. By (1), X/R is quasisimple. (Or a central extension of $SL_2(p)$, $p \leq 3$.) So if $R \neq C_X(V)$ then the p-multiplier of SL(V) is nontrivial and hence p = 2 and $X/R \cong SL_2(7)$ or $L_4(2)/Z_2$. As QR/R is abelian the first case is out. Thus (3) holds.

Assume next that $|V:V\cap Q| = p$ and $\mathcal{X}(M)$ is nonempty. For each line Z+T through Z and each $g \in M$, either $Z+T \subseteq Q^g$ or $|(Z+T)\cap Q^g| = 1$ as $Q^g\cap V$ is a hyperplane of V. Hence $\mathcal{S}(M)$ is a Shult space (also known as prepolar space) whose lines are of order p+1.

Assume the rank of the Shult space $\mathcal{S}(M)$ is at least 3. Then by the Buekenout-Shult Theorem [16], $\mathcal{S}(M)$ is the Shult space of $K = \Omega_n^{\epsilon}(p)$ or $Sp_n(p)$. Let $\bar{X} = X/C_X(V)$ and W the natural module for K. Then $\mathcal{S}(M)$ is isomorphic to the geometry of singular subspaces of W. Let $d = m(V)$ and observe $m(\bar{Q}) \geq m(Q\cap V)-1 = d-2$. By (3), if $\Gamma \subseteq \mathcal{P}(M)$ is the set of points in a maximal singular subspace of W, then $m(\langle\Gamma\rangle)$ is the rank of $\mathcal{S}(M)$. So as the rank of $\mathcal{S}(M)$ is at least 3, and as there exist disjoint maximal singular subspaces of W, we have $d \geq 6$. Thus $m(\bar{Q}) \geq 4$. Notice also that $\bar{Q}$ is an abelian subgroup of the radical $E = E(Z)$ of $N_K(Z)$ since $[Q,Q\cap V] \leq Z$ implies Q fixes each line in $\mathcal{S}(M)$ through Z.

Claim $N_{\bar{X}}(Z)$ is irreducible on the section $Z^{\perp}/Z$ of W. For if U is a maximal totally singular subspace of W then by (3) the identification of 1-dimensional subspaces of W with the corresponding conjugates of Z in V defines an isomorphism of U with a subspace V_U of V such that $SL(V_U) \leq Aut_X(V_U)$. Hence $SL(U) \leq Aut_{\bar{X}}(U)$. In particular for each such U through Z, $SL(U/Z) \le Aut_{N_{\bar{X}}(Z)}(U/Z)$. So as the graph connecting lines through Z contained in a maximal singular subspace is connected, it follows that $N_{\bar{X}}(Z)$ is transitive on lines through Z, which establishes the claim.

Next $E_0 = C_K(Z^{\perp})$ is of order at most p and the action of $N_K(Z)$ on $Z^{\perp}/Z$ is dual to that on E/E_0, so $N_{\bar{X}}(Z)$ is irreducible on E/E_0. But as $\bar{Q} \le E$ and as $m(\bar{Q}) > 1$, it follows that $E = \bar{Q}$. In particular as $\bar{Q}$ is abelian, p = 2 if K is symplectic. Also as $K = \langle E(U) : U \in \mathcal{P}\rangle$, we have $K = \bar{X}$.

Finally claim the representations of $\bar{X}$ on W and V are equivalent. As $\bar{X}$ is irreducible on V and the stabilizer of Z in each action is the same parabolic, it suffices by [21] to show the actions of a Cartan group H on Z are the same in each representation. Let $O^{p'}(N_{\bar{X}}(Z_i)) \cap H = H_i$, where $Z_0 = Z$ and Z_1 is a singular line through Z. Then $H = H_0H_1$ and the actions of H_i in each representation are the same, so the claim holds. Notice the claim completes the proof of (4) modulo the assertion that $R = C_X(V)$. The proof of this last assertion is the same as that of the corresponding statement in (3). Use the fact that QR/R is abelian to eliminate the cases $X/R \cong Sp_6(2)/Z_2$ or $\Omega_8^+(2)/E_4$. See section 21 of [12] for the necessary facts about the last two coverings.

If $[Q,V] \le Z$ then $V \le QC_G(Q)$. Thus (5) holds.

Assume either (a) or (b) of (6) holds. In case (b), $|Q:C_Q(V \cap Q)| \le p$, so $m(Q \cap V) \le 2$. If $Q \cap V = Z$ then by (5), $V = Z \le C(Q)$, a contradiction. So in either (a) or (b) we have $U = Q \cap V$ of rank 2.

Let $\tilde{Q} = Q/Z$, $H = N_G(Z)$, and $H^* = H/Q$. Then $[V,\tilde{Q}] \le \tilde{U}$, so V^* is a group of transvections on the symplectic space $\tilde{Q}$. In particular $|V^*| \le p$, so $|V| \le p^3$. If $|V| = p^2$ then $V \le Q$ and (i) holds. So let $|V| = p^3$. By (2), X is irreducible on V, so as $\bar{Q}$ is weakly closed in $N_{\bar{X}}(\bar{Q})$ and $|\bar{Q}| = p$, we conclude that p is odd and that V is the natural module for $\bar{X} \cong \Omega_3(p)$. The usual argument shows $|C_X(V):R| \le 2$, as 2 is the order of the multiplier of $\Omega_3(p)$.

(17.7) Let $M = \langle Q^M\rangle \le G$ with $V = \langle z^M\rangle \le Q$. Let $n = m(V)$ and $(z^{g_i} : 1 \le i \le n)$ a basis for V. Set $Q_i = Q^{g_i}$, $R_j = \bigcap_{i \le j} Q_i$, $1 \le j \le n$, and $\Delta = (Q_i : 1 \le i \le n)$. Let $A\alpha = A/B$ be a noncentral p-chief factor of M with $V \le B$ and define

$$m = \max\{j : R_j \cap A \not\le B\}$$

with m = 0 if $R_j \cap A \le B$ for all $1 \le j \le n$. Then $M = \langle\Delta\rangle$, $0 < m < n$, and $A\alpha$ is the m-th fundamental module for $M/C_M(V) \cong SL_n(p)$, subject to labeling the Dynkin diagram of $M/C_M(V)$ so that the k-th maximal parabolic stabilizes a k-dimensional subspace of V. In particular $A\alpha$ is of dimension $\binom{n}{m}$.

Proof. Let $z_i = z^{g_i}$. Without loss $z = z_1$. $[Q,A] \le Q\cap A$, so if $Q\cap A \le B$ then $M = \langle Q^M \rangle$ centralizes $A\alpha$, contrary to hypothesis. Thus $m > 0$.

Let $M^* = M/C_M(V)$, $U_i = \langle z_j : j \le i \rangle$, and $U = U_m$. By 17.6.3, $M^* = SL(V)$, so that $P^* = N_M(U)^*$ is a parabolic of M^* and indeed either $m = n$ and $P = M$ or P^* is the m-th maximal parabolic. Further $M^* = \langle \Delta^* \rangle$ and $C_M(V) \le N_M(Q)$, so $M = \langle \Delta \rangle$.

Let $W_0 = A$ and for $1 \le i \le n$ let $W_i = R_i \cap A$. Set $W = W_m$. For $j \le i$, $[W_i,Q_j] \le \langle z_j \rangle \le V$, so $[(W_i)\alpha,Q_j] = 0$. In particular if $m = n$ then $M = \langle \Delta \rangle$ centralizes $W\alpha$ so as M is irreducible on $A\alpha$, we have $A\alpha = W\alpha$ centralizing M, a contradiction. Thus $m < n$.

Next 17.6.3 gives the action of M on V, from which we see that $O^{p'}(P^*) = \langle (Q_{m+1}\cap P)^{*P}, Q_i^* : 1 \le i \le m \rangle$ with $\langle (Q_{m+1}\cap P)^P \rangle$ centralizing U and hence acting on W. Further $[Q_{m+1}\cap P,W] \le R_{m+1}\cap A \le B$, so $(Q_{m+1}\cap P)^*$ centralizes $W\alpha$. By the last paragraph, $(Q_i)^* \le C(W\alpha)$ for $i \le m$, and by the first sentence of this paragraph, $\langle (Q_{m+1}\cap P)^P \rangle$ acts on W. Thus $O^{p'}(P^*) = \langle (Q_{m+1}\cap P)^{*P},(Q_i)^* : 1 \le i \le n \rangle$ acts on W and centralizes $W\alpha$. Hence as P^* is a maximal subgroup of M^* and a Borel group of M^* fixes a unique point of $W\alpha$, $W\alpha$ is a point and P^* is the stabilizer in M^* of $W\alpha$.

To complete the proof we appeal to Curtis [21] and argue as in the proof of 17.6.4. In particular it suffices to show that if S^* is the standard minimal parabolic not contained in P^* and $g \in S - P$, then S stabilizes the line $W\alpha+(W\alpha)^g$. Let $X = \bigcap_{i<m} N_{Q_m}(Q_i)$, $D = W_{m-1}$, and choose $g^2 \in N(W)$. From the action of M on V, $S = \langle X,g,N_S(W)\cap N_S(W^g) \rangle$, so it remains to show $X \le N(D\alpha)$. But $[X,D\alpha] \le (Q_m\cap D)\alpha = W\alpha$, completing the proof.

(17.8) Assume $g \in G$ with $z^g \in Q - Z$, let $X = \langle Q,Q^g \rangle$, $V = \langle Z,Z^g \rangle$, $U = Q\cap Q^g$, and $R = N_Q(Z^g)N_Qg(Z)$. Then V, U, and R are normal in X, X/R acts naturally on V, X centralizes U/V, and R/U is the direct sum of natural X/R modules.

Proof. By 17.5, $V = \langle Z^X \rangle$. Then by 17.6, $R = C_X(V)$ and $X/R = SL(V)$. By 17.7, X centralizes U/V and each chief factor of X/R on R/U is natural. The semisimplicity of R/U follows from 2.5.

(17.9) Let $Q \leq T \in Syl_p(G)$, $V = J(T)$, $M = N_G(V)$, and $X = \langle Q^M \rangle$. Assume M is irreducible on $V \cong E_{p^4}$, Q is extraspecial of width 2, and there exists $g \in G$ with $Z \neq Z^g \leq Q \cap V$. Then either

(1) V is the natural module for $X/V \cong \Omega_4^+(p)$, $\mathcal{S}(M)$ is the Shult space of that orthogonal space, $m(Q \cap Q^g) = 3$, and $X = O^{p'}(M)$, or

(2) $p = 2$, $Q \cong D^2$, $Q \cap Q^g = ZZ^g$, V is the natural module for $X/V \cong Sp_4(2)'$, $\mathcal{S}(M)$ is the Shult space of that symplectic space, and X = M.

Proof. Let $M^* = M/C_M(V)$. As M is irreducible on V, $V = \langle Z^M \rangle$, so we are in the set up of 17.6. As V = J(T), M controls fusion in V, so we may take $g \in M$. In particular $\mathcal{L}(M)$ is nonempty. As V is abelian and Q is of width 2, $m(Q \cap V) \leq 3$. By 17.6.6, $m(Q \cap V) \geq 3$, so $Q \cap V$ is a hyperplane of V. Hence by 17.6.4, $\mathcal{S}(M)$ is a generalized 4-gon.

The collinearity relation on $\mathcal{P}(M)$ makes $\mathcal{P}(M)$ into a strongly regular graph with μ the number of lines through Z, $k = p\mu$, and $\lambda = p-1$. (The standard notational conventions for strongly regular graphs are used; cf section 16 of [6].) Indeed for $h \in M$ with $Z^h \not\leq Q$, μ is also the number of points in $\mathcal{P}(M)$ on the nonsingular line ZZ^h. In particular $2 \leq \mu \leq p+1$.

Suppose $\mu = p+1$. Then each 1-dimensional subspace of V is in $\mathcal{P}(M)$ and the map $Z^x \mapsto Q^x \cap V$ is an absolute polarity of the projective geometry on V preserved by G. So G preserves a symplectic form on V. If p = 2 it is easy to check that (2) holds. So take p odd. Then $E_{p^2} \cong Q^*$, X is transitive on the 1-dimensional subspaces of V, and each subgroup of Q^* of order p is in the set Δ of section 5. Hence by 5.1, 5.2, and 5.3, $X^* = Sp(V)$. As $Q^* \trianglelefteq N_{X^*}(Z)$ this is impossible.

So assume $\mu \leq p$. Let $Y = \langle Q, Q^g \rangle$ and $U = Q \cap Q^g$. By 17.8, $U = ZZ^g$ or U is of rank 3. Assume the former. Then from 17.8, $Y^* \cong SL_2(p)/E_{p^2}$. Now $O_p(Y^*)$ is transitive on the p 2-dimensional subspaces in $Q \cap V$ through Z distinct from ZZ^g, so as at least one of those subspaces is in $\mathcal{L}(M)$, all are. But then $\mu = p+1$, contrary to assumption.

Thus U is of rank 3. In this case $Y^* \cong Z_p \times SL_2(p)$. Claim $\mu = 2$. Let y^* be of order p-1 in Y^* and acting on Z and Z^g. Then y^* has two eigenvalues λ and λ^{-1}

on U, where λ generates $F^{\#}$. As y^* commutes with U^*, the action of y^* on $(Q\cap V)/Z$ is the same. A similar element y_L^* exists for each $L \in \mathcal{L}(M)$ with $Z \leq L$. Let $B^* = \langle y_L^* : Z \leq L \in \mathcal{L}(M)\rangle$. From the subgroup structure of $SL_2(p)$, one of the following holds:

(i) B^* is transitive on the 1-dimensional subspaces of $(Q\cap V)/Z$.

(ii) $\mu = 2$ or p.

(iii) p = 5 or 7 and $\mu = (p+1)/2$.

As $\mu \neq p+1$, case (i) is out. A check of the numerical restrictions on the parameters of $\mathcal{S}(M)$ (cf. 16.7 in [6]) show $\mu \neq p$ or $(p+1)/2$. So $\mu = 2$ as claimed.

Now a moment's thought shows $\mathcal{S}(M)$ is the Shult space whose point set is $\Delta \times \Delta$, where Δ is the projective line over GF(p), whose lines consist of the subsets $L_a = \{(x,a) : x \in \Delta\}$ and $L^a = \{(a,x) : x \in \Delta\}$, $a \in \Delta$, and with incidence defined by inclusion. But this is just the Shult space of an orthogonal form of sign +1 on V, so $X^* \leq O_4^+(p)$. Indeed as X is irreducible on V and generated by conjugates of $Q^* \cong E_{p^2}$, $X^* \cong \Omega_4^+(p)$. Hence (1) holds.

18. Almost simple groups of GF(p)-type.

In this section we determine the almost simple groups of GF(p)-type. The proof assumes the Classification Theorem. Alternatively the result is applicable to K-groups without the Classification.

Theorem A. Let p be a prime, G a finite group with a large symplectic p-subgroup Q, $Z = \Omega_1(Z(Q))$, and $L = \langle Z^G \rangle$. Assume $G = LQ$ with L a nonabelian simple group. Then one of the following holds: (And indeed each is an example.)

(1) $G = L$ is of Lie type over GF(p), $Q = F^*(C_G(Z))$, and Z is a long root subgroup of G. G is not a Ree group, $F_4(2^n)$, $Sp_4(2^n)$, or $G_2(3)$.

(2) p = 2 and either Q is of width 1 or $G \cong A_9$, A_{10}, or $E_{2k}/P\Omega_6^-(3)$, k = 1 or 2.

(3) p = 2, $G \cong Aut(M(22))$, and $Q = O_2(C_G(Z)) \cong D^5$.

(4) p = 2, $G \cong M_{22}$, and $Q = O_2(O^2(C_G(Z))) \cong D^2$.

(5) p = 2, $G \cong Aut(HS)$, and $Q = F^*(C_G(Z)) \cong D^3$.

(6) p = 3, $Q = O_3(C_G(Z)) \cong 3^{1+4}$, and G is $\Omega_8^+(2)$ extended by a triality automorphism.

(7) p = 3, $Q \cong 3^{1+2}$, and $G \cong A_n$, $9 \le n \le 11$, $PGL_3(4)$, or $U_3(8)$.

(8) p = 5, $Q \cong 5^{1+2}$, and $G \cong U_5(4)$.

(9) $G = L$, $Q = F^*(C_G(Z))$, and G and the order of Q (or the isomorphism type of Q if p = 2) are listed in Table A.

Throughout this section assume the hypothesis and notation of Theorem A. Also let $\langle z \rangle = Z$, $Q \le T \in Syl_p(G)$, $H = N_G(Z)$, and $P = O_p(H)$. If p = 2 assume Q has width at least 2. This does no harm as the alternative is one of the conclusions of Theorem A.

To see that the examples listed satisfy hypothesis L4 use 17.2 and 17.3. For example if Z is not weakly closed in Q with respect to G then 17.5 (Whose proof does not use L4.) and 17.3 show L4 is satisfied. Moreover we'll see in 19.4 and 21.2 that usually Z is not weakly closed in Q. Verification of the other axioms will emerge during the proof of Theorem A.

From hypothesis L3 in the definition of large symplectic subgroups, $Q \le P$. So

Table A

G	Q	G	Q	G	Q
$G_2(3)$	D^2	$P\Omega_6^{\epsilon}(3)$	D^2	$P\Omega_8^{+}(3)$	D^4
${}^3D_4(2)$	3^3	${}^2F_4(2)'$	3^3	$Sp_6(2)$	3^3
$U_6(2)$	3^5	$F_4(2)$	3^5	${}^2E_6(2)$	3^7
M_{12}	$D^2,3^3$	J_2	QD	J_3	QD
J_4	$11^3,D^6$	HS	$Z_4*D^2,5^3$	Mc	$3^5,5^3$
He	$D^3,7^3$	Ly	5^5	Ru	5^3
Co_1	$D^4,3^5,5^3$	Co_2	$D^4,3^5,5^3$	Co_3	$3^5,5^3$
ON	7^3	Sz	Q^3	M_{24}	D^3
M(22)	3^7	M(23)	3^9	M(24)′	$D^6,3^{11},7^3$
F_5	$D^4,3^5,5^5$			F_2	$D^{11},3^9,5^5$
F_1	$D^{12},3^{13},5^7,7^5,13^3$			F_3	$D^4,5^3$

by 17.1:

(18.1) (1) Q is weakly closed in T with respect to G.

(2) $Z \le Z(T) \le Z(P) \le Z(Q)$, and each of these subgroups is cyclic.

Each is equal to Z if Q is extraspecial.

(18.2) P is nonabelian and if $p = 2$ then P is of sectional 2-rank at least 4.

Proof. This is because $Q \le P$.

Let's first consider the case $L \in Chev(p)$. We appeal to section 10. In particular by 10.1, either $Z(T)$ contains a long root element or $L \cong F_4(q)$ or $Sp_4(q)$ and some element of T acts nontrivially on the Dynkin diagram of L, or $L \cong Sp_4(2)'$. (If $p = 3$ and L is $G_2(q)$ then $Z(T)$ contains long and short root elements, and hence is noncyclic.) We handle the exceptional cases first.

As $Aut(A_6)$ has sectional 2-rank 3, L is not A_6 by 18.2. In the remaining cases $Z(T)$ contains the product of a long and a short root element, so by 18.1, z is such a product. Now H is described in [12]. In particular $P = O_2(H \cap L)$, so as $Q \le P$ and $G = LQ$, we have $G = L$. This contradicts the assumption that T is nontrivial on the Dynkin diagram of L.

Therefore T is trivial on the Dynkin diagram of L if L is $F_4(q)$ or $Sp_4(q)$ and by 18.1, z is a long root element of L. This shows L is not $F_4(q)$ or $Sp_4(q)$ when $p = 2$, since in those cases $Z(T)$ is noncyclic. By 10.2, one of the following holds:

(a) $P = F^*(H \cap L)$.

(b) $p = 2$, $G \cong S_8$, ${}^2F_4(2)$, $G_2(2)$, $PGL_2(7)$, or $L_3(4)$ extended by a graph automrophism and $|P:O_2(H \cap L)| = 2$.

(c) $p = 3$, $G = {}^2G_2(3)$.

Again we handle the exceptional cases first. $PGL_2(7)$ is of sectional 2-rank 2, so that case is out by 18.2. If G is ${}^2G_2(3)$ then $\Omega_1(T) \cong E_9$, whereas Q is a nonabelian subgroup of $\Omega_1(T)$. If $G = G_2(2)$ then $P \cong D_8^2$, which forces $P = Q$, so that case (1) of Theorem A holds. If G is $Z_2/L_3(4)$ then $Z(P)$ is noncyclic.

To handle the remaining cases the following observation is useful. Let X be of odd prime order in H. $Z = C_{P \cap L}(X)$ so $P = (P \cap L)C_P(X)$ with $C_P(X) = C_Q(X)$. Observe that $C_P(X)$ is of order 4 and not normal in H. However in general we have:

(18.3) If X is a p'-subgroup of H then $C_Q(X) = Z(Q)*Q_X$ with $Q_X = 1$ or extraspecial, $[Q,X]$ is extraspecial, and $[Q,X,C_Q(X)] = 1$.

In particular returning to our treatment of the exceptional cases, as $C_Q(X)$ is of order 4, 18.3 says $C_Q(X) = Z(Q) \trianglelefteq H$, contradicting $C_Q(X)$ not normal in H.

We are left with the case $P = F^*(H\cap L)$. Hence as $G = LQ$ with $Q \le P$, we have $G = L$. Therefore the root group R of z is contained in $Z(T)$, so $R = Z$ by 18.1. Hence R is of order p, so L is defined over $GF(p)$.

If L is the Tits group there is a unique normal subgroup of H between P and Z and that is abelian. Thus $P = Q$, which is impossible as P is not of symplectic type. If L is $G_2(3)$ then $Z(T)$ is noncyclic while ${}^2G_2(3)' \cong L_2(8)$ has abelian Sylow 3-groups. In either case we have a contradiction by 18.1 or 18.2.

For all other choices of L defined over $GF(p)$, 10.2 and 10.3 say either H is irreducible on P/Z or $L \cong L_n(p)$ and there are just two H-invariant subgroups between Z and P, both of which are abelian, or $L \cong L_3(2)$ and all proper subgroups of P are abelian. We conclude $P = Q$. Then as P is extraspecial, Theorem A is established in this case.

We have shown:

(18.4) If $L \in Chev(p)$ then Theorem A holds.

Thus from now on we assume $L \notin Chev(p)$.

We next assume $L \cong A_n$. Represent L on $\Omega = \{1,\dots,n\}$ and write $n = \sum_{i=0}^{\infty} a_i p^i$ in its p-ary expansion. By 7.3, T has a_i orbits Ω_{ij}, $1 \le j \le a_i$, of length p^i on Ω, and either $Z(T)$ is of p-rank $m = \sum_{i=1}^{\infty} a_i$ or $p = 2$, $L = G$, and $m(Z(T)) \ge m-\epsilon$, where $\epsilon = 1$ if $a_1 = 1$ and $\epsilon = 0$ otherwise. We conclude from this observation and 18.1 that $n = p^i+k$ for some $i > 1$ and some $0 \le k \le p+1$. Moreover $k < p$ unless $p = 2$ and $G = L$. Further the set Δ of cycles of z is of order $d = p^{k-1}$. But now $P = G_\Delta$ is abelian unless $p = 2$ and $i \le 3$ or $p = 3$ and $i = 2$, so one of the exceptional cases must hold by 18.1.

If $p = 2$ then $n \ne 5,6$, or 8 by 18.4. This leaves $n = 7,9,10$, or 11, with $G = L$ unless possibly $n = 9$. Actually if $n = 9$ we must have $G = L$ since if $G \cong S_9$ then H

is contained in an S_8 subgroup, a case already eliminated. So it remains to observe n cannot be 7 or 11 since there $O_3(H) \neq 1$.

If $p = 3$ then $n = 9,10$, or 11. In this case $T \cong Z_3wrZ_3$, so $Q \cong 3^{1+2}$ and Theorem A holds.

We have shown:

(18.5) If $L \cong A_n$ then Theorem A holds.

So in the remainder of this section we assume L is not A_n.

Assume next that L is a sporadic group. Then Out(L) is of order at most 2 so as $G = LQ$ we conclude either $G = L$ or $p = 2$. The conjugacy classes of subgroups of order p and their centralizers in G are listed in section 5 of [23]. As T is nonabelian, Z(T) is cyclic with $Z \leq Z(T)$, and $F^*(H) = O_p(H)$, we conclude from this list that:

(18.6) If L is a sporadic group then one of the following holds:

(1) P is of symplectic type, $P \leq L$, and L and the order (or isomorphism type) of P is listed in Table A.

(2) $p = 3$ and $G \cong J_3$ or F_3.

(3) $p = 2$ and $L \cong M_{11}, M_{12}, M_{22}, M_{23}$, HS, Ru, or M(22), and if $L \cong M_{12}$ or HS then $L \neq G$.

Suppose 18.6.2 holds. If $L \cong J_3$ then $\Omega_1(T) \cong E_{27}$, whereas $Q = \Omega_1(Q)$ is nonabelian, a contradiction. If $L \cong F_3$ then 14.3 supplies a contradiction; namely as $Q \trianglelefteq H$ and $Q \neq Z$, $D_3 \leq Q$. (In the notation of 14.3.) Similarly $Q \not\leq C_G(D_3) = C_3$ and then by 14.3.3, $O_3(H) = Q\Phi(O_3(H))$, so $Q = O_3(H)$. However by 14.3, $O_3(H)$ has class greater than 2.

Suppose 18.6.3 holds. By 18.2 L is not M_{11}. If $G \cong Aut(M(22))$ then $P \cong D^5$ and we add G to the list of 18.6.1 and complete the case later. G is not M_{23} or M(22) since in those cases Z(P) is noncyclic.

Suppose L is M_{12} or M_{22}, let X be of order 3 in H, and let $R = [P,X]$. Then $R \cong Q^2$ and $C_P(X) \cong Z_4$, E_4, or D_8 for $G \cong Aut(M_{12})$, M_{22}, or $Aut(M_{22})$, respectively. In each case every proper H-invariant subgroup of R is abelian, so 18.3 implies $R \leq Q$. Next $C_P(XR) \cong Z_2$, Z_2, or E_4, respectively, so 18.3 eliminates the third case and says $Q = R$ in the first two cases. In particular $Q \leq L$ so $G = L$ and hence G is M_{22} and Theorem A holds.

If G is Aut(HS) then $P \cong D^3$ and the only H-invariant proper subgroups of P are $P \cap L \cong Z_4 * D^2$ and the subgroups of $Z(H \cap P)$. Thus Theorem A holds in this case and also when $G \cong HS$.

This leaves $G \cong Ru$, where we appeal to 12.12, which says $P/\Phi(P) \cong E_{64}$, H is indecomposable on $P/\Phi(P)$ with $C_{P/\Phi(P)}(H) = P_1/\Phi(P) \cong E_4$, H irreducible on P/P_1, and $P_1 \cong Q_8 \times E_{16}$. If $Q \not\leq P_1$ then as H is irreducible on P/P_1 and indecomposible on $P/\Phi(P)$, we must have $P = Q$. This is impossible as P is of class 3. So $Q \leq P_1$, contradicting $C_G(Q)$ cyclic.

So when L is sporadic it remains to consider 18.6.1 less the case $G = HS$ but with the case $G \cong Aut(M(22))$ added. Now $Q \leq P$ with P extraspecial, so if Q is extraspecial then $P = Q*C_P(Q)$. Then as $C_G(Q) \leq Q$, $P = Q$ and we are done. If Q is not extraspecial then $p = 2$ and at least $|P:QC_P(Q)| \leq 2$, so $|P:Q| \leq 2$. However there is no H-invariant subgroup of P of index 2 unless $L = M(22)$, where the unique such subgroup is not of symplectic type.

We have shown:

(18.7) If L is sporadic then Theorem A holds.

In particular we have reduced to the case where $L \in Chev(r)$ for some prime $r \neq p$. First consider the case where $p = 2$. Then in the terminology of section 9, $\Delta = Fun(T)$ is nonempty. Let $X = N_G(\Delta)$. Then $T \leq X$ and $O_2(X) \neq 1$, so by 17.4, $F^*(X) = O_2(X)$. But choosing $K \in \Delta$, $K \trianglelefteq \trianglelefteq X$ and $K \cong Sl_2(q)$, where L is defined over GF(q) or $L \cong {}^3D_4(q^{1/3})$. Thus $q = 3$ and L is not ${}^3D_4(3)$.

Now $SL_2(3) \cong K \trianglelefteq \trianglelefteq X$ so K centralizes every abelian normal subgroup of X. In particular K centralizes $\langle Z^X \rangle$. Hence $K \trianglelefteq \trianglelefteq C_X(Z)$ so $K \trianglelefteq KP$. Therefore $Z(K) \leq C_G(Q) = Z(Q)$, so $Z = Z(K)$. As this holds for each $K \in \Delta$, it follows from Theorems 6 and 8 in [10] that $L \cong L_3^{\epsilon}(3)$, $G_2(3)$, or $P\Omega_n^{\epsilon}(3)$, $5 \leq n \leq 8$. As $U_3(3) = G_2(2)'$ and $P\Omega_5(3) \cong U_4(2)$ are in Chev(2), L is neither of these groups.

As $K \trianglelefteq KQ$ and $C_G(Q) \leq Q$, we have $O_2(K) \leq Q$. Thus $R = \langle O_2(K) : K \in \Delta \rangle \leq Q$. If L is $G_2(3)$ or $P\Omega_8^+(3)$, or if L is $P\Omega_6^{\epsilon}(3)$ and $G = L$ then $R = P$, so $P = Q$. Thus Theorem A holds in these cases.

Lemma 18.2 shows L is not $L_3^{\epsilon}(3)$. If L is $P\Omega_8^-(3)$ then $F^*(H)$ is not a 2-group while if L is $P\Omega_7(3)$ there is a normal A_4-subgroup of H which must then centralize Q. In either case we have a contradiction, so we are left with $L \cong P\Omega_6^{\epsilon}(3)$ and $L \neq G$. Theorem A holds by inspection there.

Therefore p is odd. Now z is a semisimple element of L and hence is contained in a maximal torus X of the subgroup D of G inducing inner-diagonal automorphisms on L. We appeal to 9.1 to conclude there are normal subgroups Y_0 and $Y = XY_0$ of $C_D(Z)$ such that $Y_0 = D_0E$ where $E = C_Y(D_0) \leq X$ and $D_0 = D_1 \ldots D_k$ is a central product of groups of Lie type defined over $GF(q^{e_i})$, where L is defined over GF(q).

As $F^*(N_G(Z)) = P$, either $D_0 = 1$ or $q = 2$, $p = 3$, and $D_i \cong L_2(2)$, $U_3(2)$, or $SU_3(2)$ for each i, $1 \leq i \leq k$. As $Q \leq P$, Q acts on D_i. Thus if $D_i \neq 1$ then $1 \neq C_{D_i}(Q) \leq Z$ and hence $Z \leq Z(D_i)$. It follows that $D_i \cong SU_3(2)$ and $Z = Z(D_i)$. Further either $O_3(D_i) \leq Q$ or $[D_i,Q] = 1$, and the latter contradicts $C_G(Q) \leq Q$. Thus we have shown:

(18.8) Either

(1) $X = Y \trianglelefteq C_G(Z)$, or

(2) $q = 2$, $p = 3$, and $D_0 = D_1 \ldots D_k$ is the central product of $k \geq 1$ copies of $SU_3(2)$ with $Z = Z(D_i)$ and $O_3(D_0) \leq Q$.

Next from the 3-local structure of groups over GF(2) we conclude:

(18.9) If 18.8.2 holds then one of the following holds:

(1) $P \leq L$, P is extraspecial, and L and $|P|$ are described in Table A.

(2) $L \cong \Omega_8^{\epsilon}(2)$ or $U_9(2)$.

If 18.9.1 holds we conclude as above that $P = Q$ and hence Theorem A holds. On the other hand if 18.9.2 holds then either $|Z(P)| = 9$ or G is $\Omega_8^+(2)$ extended by a triality automorphism and $P \cong 3^{1+4}$. The first case contradicts 18.1 and in the second Theorem A holds.

Thus we may assume 18.8.1 holds. By 9.5:

(18.10) Either G induces inner diagonal automorphisms on L or $p = 3$, $L \cong D_4(q)$ or ${}^3D_4(q)$, and $|G:D| = 3$ with some element of Q inducing an automorphism in the coset of a graph or graph-field automorphism.

Further by 9.1 and 9.2:

(18.11) $(Q\cap D)Y/Y$ is of order at most p with equality only if $L \cong L_n^{\epsilon}(q)$ with $(q-\epsilon,n) \equiv 0 \bmod p$ or $p=3$, $L \cong E_6^{\epsilon}(q)$, and $q \equiv \epsilon \bmod 3$.

Let $p^a = |G:D|$ and $p^b = |(Q\cap D)Y:Y|$. Observe that not both a and b can be 1 so $|Q:Q\cap Y| = p^{a+b} \le p$. Therefore as $Q\cap Y$ is abelian, Q is of width 1 and $a+b = 1$. Also as $F^*(C_G(Z))$ is a p-group, so is Y. Further $Q = C_G(Q/Z)$ while $[Y,Q/Z] \le (Q\cap Y)/Z$, so $|Y:Q\cap Y| \le p$. Thus

(18.12) $|G|_p = p|Y| \le p^4$.

Suppose $a = 1$. Then $p = 3$ and G is described in 18.10. But then $|L|_3 > 3^3$, contrary to 18.12. Therefore $b = 1$ and $G = D$ is described in 18.11. Now 18.12 implies either $p = 3$ and $G \cong PGL_3(4)$ or $U_3(8)$ or $p = 5$ and $G \cong U_5(4)$. In particular Theorem A holds, so the proof is complete.

19. Random lemmas on groups of GF(p)-type.

In this section assume Q is a large symplectic p-subgroup of the finite group G with Q of width at least 2, $L = F^*(G)$ simple, $G = LQ$, and L not of Lie type and characteristic p. Exclude case (2) of Theorem A. Let $Z = \Omega_1(Z(Q)) = \langle z \rangle$, $H = N_G(Z)$, $H^* = H/Q$, and $\tilde{H} = H/Z$.

(19.1) One of the following holds:

(1) H is irreducible on $\tilde{Q}$.

(2) p = 2 and G is $G_2(3)$, M_{12}, or M_{22}.

(3) p = 2, $L \cong HS$ or $M(22)$, and $Z(Q\cap L)/Z$ and $Q\cap L$ are the only proper nontrivial H-submodules of $\tilde{Q}$.

(4) p = 2, $G \cong M_{24}$ or He, and there are just two proper nontrivial H-submodules of $\tilde{Q}$, each totally singular of rank 3.

(5) p = 3 and $G \cong Z_3/\Omega_8^+(2)$.

Proof. By Theorem A we know G and z and hence also the general structure of H. For example if L is sporadic this information appears in section 5 of [23]. Or look in the tables in the introduction of this paper. Usually it's easy to see the minimal degree of a faithful $GF(p)H^*$-module is $m(\tilde{Q})$, so that H is irreducible on $\tilde{Q}$.

Perhaps a few words should be said when p = 3 and G is M(22) or ${}^2E_6(2)$. In that event $Q \cong 3^{1+6}$. If G is ${}^2E_6(2)$ then $C_G(Z) = H_0E$ with $E \cong E_9$, $H_1*H_2*H_3 = H_0 \trianglelefteq H$, where $H_i \cong SU_3(2)$ and E is transitive on the subgroups H_i, $1 \le i \le 3$. So if G is ${}^2E_6(2)$ then evidently H is irreducible on $\tilde{Q}$. Also from the discussion in section 15, $M(22) \cong X \le G \cong {}^2E_6(2)$ with $Q \le X$ and $|C_X(Z)/Q| = 9.2^7$. In particular we may take $E \le C_X(Z)$, so as each subgroup of $C_G(Z)$ of order $|C_X(Z)|$ containing E is irreducible on $\tilde{Q}$, we are alright in this case too.

(19.2) Define

$$m(H^*,\tilde{Q}) = \min\{m([\tilde{Q},h^*]) : h^* \text{ of order } p \text{ in } H^*\}.$$

and assume H^* is of order divisible by p and $m = m(H^*,\tilde{Q}) \le 2$. Then one of the following holds:

(1) $Q \cong 3^{1+4}$, $m = 2$, and $G \cong F_5, Mc, Co_3, Co_2, Z_3/\Omega_8^+(2), U_6(2)$, or $F_4(2)$.

(2) $Q \cong 3^{1+4}$, $m = 1$, and $G \cong Co_1$.

(3) $Q \cong QD$ or D^2, $m = 2$, and $G \cong G_2(3), P\Omega_6^\epsilon(3), M_{12}, M_{22}, J_2$, or J_3.

(4) $Q \cong D^3$ or Q^3, $m = 2$, and $G \cong M_{24}$, He, or Sz.

(5) $Q \cong Z_4{*}D^2$ or D^3, $m = 2$, and $L \cong HS$.

(6) $Q \cong D^4$, $m = 2$, and $G \cong Co_1$ or Co_2.

(7) $Q \cong D^5$, $m = 2$, and $G \cong Aut(M(22))$.

Proof. The commutator map on Q defines a symplectic form on $\tilde{Q}$ which is nondegenerate if Q is extraspecial and which is preserved by H^* up to a scalar. Also we know Q and H^*. Thus if p is odd we can appeal to 19.1 and Theorems 5.1 and 5.2. If $p = 2$ inspect the action of H^* on $\tilde{Q}$ or appeal to 12.16 in Timmesfeld [37].

(19.3) Let $G \cong Co_1$, $Q \cong 3^{1+4}$, $Q \le T \in Syl_3(G)$, and $J = J(T)$. Then

(1) G has four classes of subgroups of order 3 with representatives Z, B, C, and D.

(2) $N_G(Z) \cong GPSp_4(3)/3^{1+4}$.

(3) $N_G(B) \cong Z_2/Sz/Z_3$.

(4) $N_G(C) \cong E_4/U_4(3)/E_9$.

(5) $N_G(D) \cong S_3 \times A_9$.

(6) $J \cong E_{3^6}$ and $N_G(J)/J \cong M_{12}/Z_2$.

(7) We may choose the representatives Z, B, C to be contained in J. Moreover

(i) $|Z^{N(J)}| = 220$ and $N_H(J)/J \cong (GL_2(3)/E_9)xZ_2$.

(ii) $|B^{N(J)}| = 12$ and $(N(J)\cap N(B))/J \cong Z_2 \times M_{11}$.

(iii) $|C^{N(J)}| = 132$ and $(N(J)\cap N(C))/J \cong Z_2/(Z_2 \times A_6)$.

Proof. The first five remarks can be found in section 5 of [23]. We conclude from (4) there is $E_{3^6} \cong J \le T$. From (2) we conclude $I = J\cap Q$ is the unique T-invariant subgroup of Q of order 27 and $J^* = C_{H^*}(\tilde{I}) = J(T^*)$. But then as $I = C_Q(I)$ we conclude $J = C_G(I)$ is the unique member of $\mathfrak{A}(T)$. Therefore $J = J(T)$. In particular $N_G(J)$ controls fusion in J.

Let $M = N_G(J)$ and $\bar{M} = M/J$. We have just determined $H\cap M$ and from (3),(4), (5), and

11.4 we determine $N_M(B)$ and $N_M(C)$ and see that $D^G \cap J$ is empty. Indeed from 11.4 we know $N_M(B)$ has two orbits on the points of J/B of length 11 and 110. Let U_i, i = 1,2, be representatives of lines in the respective orbits. The centralizer of U_1 is nonsolvable, while $N_M(Z)$ is solvable, so U_1 contains no conjugate of Z. $|\bar{M}|_3 = 27$ so $|Z^M|$ is prime to 3 while $|B^M|$ and $|C^M|$ are divisible by 3 but not 9. From this information we determine Z^M consists of 220 points fused into U_2, B^M is of order 1+m, for m = 11 or 110, and the remaining points constitute C^M. This allows us to calculate that the order of $\bar{M}$ is equal to that of M_{12}/Z_2. So as 37 does not divide that order, m = 11. Now M is 2-transitive on B^M of order 12 with M_{11} the stabilizer of B, so M induces M_{12} on B^M. This completes the proof of the lemma.

(19.4) Let $T \in Syl_p(H)$ and $\mathfrak{X}(T)$ the set of subgroups X of Q of order p and distinct from Z with $\tilde{X} \le Z(\tilde{T})$. Then one of the following holds:

(1) $\mathfrak{X}(T) \subseteq Z^G$.

(2) $Q \cong 3^{1+4}$, $G \cong Co_1$, and $N_G(X) \cong E_4/U_4(3)/E_9$.

(3) $Q \cong 3^{1+4}$, $G \cong Co_2$, and $N_G(X) \cong S_3 \times Aut(PSp_4(3))$.

(4) $Q \cong 3^{1+4}$, $G \cong Co_3$, and $N_G(X) \cong (Z_2 \times S_5)/E_{3^5}$.

(5) $Q \cong 3^{1+4}$, $G \cong Mc$, and $N_G(X) \cong S_4/E_{3^4}$.

(6) $Q \cong 3^{1+4}$, $G \cong U_6(2)$, and $N_G(X) \cong Z_2/(Z_3 \times U_4(2))$ or $(Z_2 \times S_4)/E_{81}$.

(7) $Q \cong 3^{1+4}$, $G \cong F_4(2)$, and either $X \in Z^G$ or $N_G(X) \cong S_3 \times Sp_6(2)$.

(8) $Q \cong 3^{1+4}$, $G \cong F_5$, and either $X \in Z^G$ or $N_G(X) \cong Z_2/(Z_3 \times A_9)$.

(9) $Q \cong 3^{1+4}$, $G \cong Z_3/\Omega_8^+(2)$, and either $X \in Z^G$ or $N_G(X) \cong Z_2/(Z_3 \times U_4(2))$.

(10) $Q \cong D^2$, $G \cong L_4(3)$, and $N_G(X) \cong E_4/(Z_2 \times A_6)$.

(11) $Q \cong D^3$, $G \cong He$, and either $X \in Z^G$ or $N_G(X) \cong Z_2/L_3(4)/E_4$.

(12) $Q \cong D^3$, $G \cong M_{24}$, and either $X \in Z^G$ or $N_G(X) \cong S_5/E_{64}$.

(13) $Q \cong D^4$, $G \cong Co_2$, and $N_G(X) \cong A_8/(E_{16} \times D^3)$.

(14) $Q \cong D^5$, $G \cong Aut(M(22))$, and either $X \in Z^G$ or $N_G(X) \cong Z_2/U_6(2)/Z_2$ or a group of characteristic 2 and order $2^{17}.3^3$.

Proof. Let $X \in \mathfrak{X}(T)$. Then $|T:C_T(X)| = p$, so either $X \in Z^G$ or $C_T(X) \in Syl_p(C_G(X))$. By inspection of the centralizers of elements of order p in the groups listed in Theorem A, one of the following holds:

(a) $|C_G(X)|_p < |T|/p$ for all subgroups of G of order p not in Z^G.

(b) G,Q, and $N_G(X)$ are listed in Lemma 19.4.

(c) $G \cong Co_1$ and $Q \cong 3^{1+4}$.

In particular if X does not satisfy one of the conclusions of the lemma then case (b) or (c) must hold, and in case (b) we must show Z is weakly closed in Q unless $X \in Z^G$ is allowed in the appropriate case of 19.4.

Let's first consider the case $G \cong Co_1$. From the action of .0 on the Leech lattice we have $Q \le M \le G$ with $M \cong Co_2$. There is $Y \le M$ with $C_M(Y) \cong Z_3 \times Aut(PSp_4(3))$. Let $S \in Syl_3(M)$ with $Q \le S$. Without loss $C_S(Y) \in Syl_3(C_M(Y))$. Then $|S:C_S(Y)| = 3$ so $Y \le Q$ since by 19.2 no member of M induces a transvection on $\tilde{Q}$. Then as $O_3(C_M(Y)) = Y$, $Y \notin Z^G$ by 17.5. So by 19.3 we may take Y = B or C. As H is transitive on $\tilde{Q}^{\#}$ we may take X = Y. From 19.3, $N_H(J)$ is transitive on the 12 subgroups of order 3 in $(Q \cap J) - Z$, so as $N_G(J)$ is irreducible on J and $|B^{N(J)}| = 12$, X = C. Therefore 19.4 is established in this case.

Notice we have also shown:

(19.5) If $G \cong Co_1$ and $Q \cong 3^{1+4}$ then Z is weakly closed in Q with respect to G.

If p = 2 and $G = L_4(3)$ then Z is weakly closed in Q. So to complete the proof of lemma 19.4 we appeal to the following lemma:

(19.6) Let $3^{1+4} \cong Q \le T \in Syl_3(G)$ with G not Co_1 or Co_3. ($G \in Chev(3)$ is allowed.) Then $|T| = 3^6$, $V = J(T) \cong E_{81}$, and setting $X = O^{3'}(N_G(V))$ either

(1) $G \cong Z_3/\Omega_8^+(2), F_4(2), F_5$, or $L_4(3)$, V is the natural module for $X/V \cong \Omega_4^+(3)$, and Z is not weakly closed in $Q \cap V$.

(2) $G \cong U_6(2), Co_2, Mc$, or $U_4(3)$, V is the natural module for $X/V \cong \Omega_4^-(3)$, and Z is weakly closed in Q.

From the action of .0 on the Leech lattice we know that Co_3 and the groups G listed in 19.6.2 are contained in Co_1, and each contains $K \cong U_4(3)$. Thus Z is weakly closed in Q for these groups by 19.5. Notice this completes the proof of 19.4. Also setting $Y = O^{3'}(N_K(V))$, we see that V is the natural module for $Y/V \cong \Omega_4^-(3)$. Let $\Delta = Z^Y$. Then each subgroup of V of order 3 not in Δ is fused into Q under Y, so as Z is weakly closed in Q, $\Delta = Z^X$. Hence X is 2-transitive on Δ with Q^{Δ} regular on $\Delta - \{Z\}$, from which we conclude $X^{\Delta} = L_2(9) = Y^{\Delta}$, and then that X = Y. So the proof of

19.6.2 is complete.

Next if G is $Z_3/\Omega_8^+(2)$ or $L_4(3)$ its easy to see $N_G(V)$ is irreducible on V and Z is not weakly closed in $Q\cap V$. Further $F_4(2)$ contains a $Z_3/\Omega_8^+(2)$ subgroup and F_5 contains an A_{12} subgroup by 15.12, so the same remark holds for those groups. Now 17.9 completes the proof of 19.6.1.

The next lemma is essentially 3.9 and 3.10 from [37]. A proof is included for completeness.

(19.7) Assume $m(H^*,\tilde{Q}) > 2$ and $Z \neq Z^g \leq Q$. Let w be the width of Q, $P = Q^g \cap H$, and $X = Z^g$. Then

(1) $C_{\tilde{Q}}(P^*) = \tilde{X}$.

(2) $m(Q\cap Q^g) = w+1$, $Q \cong D^w$, and $P^* \cong E_{p^{w-1}}$.

Proof. $P = Z \times R$ with R symplectic. Notice $\tilde{P} \cong R$, from which it follows that $\tilde{X} = C_{\tilde{U}}(P)$, where $U = Q\cap Q^g$. So if (1) fails then P centralizes $\tilde{Y}$ of order p in $\tilde{Q}$ but not in $\tilde{U}$. Y acts on P and then also on $[P,P] = X$. Further $[P,Y] \leq Z$ so Y centralizes a subgroup of index at most p in P and then as P is of index p in Q^g, Y centralizes a subgroup of index at most p^2 in Q^g. But as $Y \not\leq U$, $Y \not\leq Q^g$, contradicting the hypothesis that $m(H^*,\tilde{Q}) > 2$. Therefore (1) is established.

Next U is elementary abelian so $m(U) = m \leq w+1$. Thus $m(P^*) = 2w+\epsilon-m \geq w+\epsilon-1$, where $\epsilon = 0$ if Q is extraspecial and 1 otherwise. So if (2) fails then $A^* = C_{P^*}(U) \neq 1$. $[C_Q(X),P,A] \leq [U,A] = 1$, so as P^* is abelian the 3-subgroup lemma shows $[\tilde{V},A^*,P^*] = 1$, where $V = C_Q(X)$. Hence by (1), $[\tilde{V},A^*] \leq X^*$. So as $\tilde{V}$ is of codimension 1 in $\tilde{Q}$, $m([\tilde{Q},a^*]) \leq 2$ for each a^* in $(A^*)^\#$, contradicting $m(H^*,\tilde{Q}) > 2$.

For $T \in Syl_p(H)$ define $\mathcal{Y}(T)$ to consist of those conjugates Z^g of Z with $Z \neq Z^g \leq Q$ and $Q^g\cap H \leq T$.

(19.8) Assume G is not M_{22} and let $T \in Syl_p(H)$. Then $\mathcal{Y}(T) \subseteq \mathfrak{X}(T)$ and one of the following holds:

(1) $\mathcal{Y}(T) = \mathfrak{X}(T)$ is of order p and Q is transitive on $\mathcal{Y}(T)$.

(2) $Q \cong 3^{1+4}$, $G \cong Z_3/\Omega_8^+(2)$, F_5, or $F_4(2)$, $\mathcal{Y}(T)$ is of order 6, and Q has two orbits on $\mathcal{Y}(T)$.

(3) $Q \cong 3^{1+4}$, $G \cong Co_1, Co_2, Co_3$, Mc, or $U_6(2)$, and $\mathcal{Y}(T)$ is empty.

(4) $Q \cong D^3$, $G \cong M_{24}$ or He, $\mathcal{Y}(T)$ is of order 4, and Q has two orbits on $\mathcal{Y}(T)$.

(5) $p = 2$, $G \cong Co_2$ or $L_4(3)$, and $\mathcal{Y}(T)$ is empty.

(6) $Q \cong D^5$, $G \cong Aut(M(22))$, $\mathcal{Y}(T)$ is of order 2, and Q is transitive on $\mathcal{Y}(T)$.

Proof. If Z is not weakly closed in Q and $m(H^*,\tilde{Q}) > 2$ the lemma follows from 19.7.1, so assume otherwise. By 19.4 - 19.6, Z is weakly closed in Q precisely when (3) or (5) holds. So assume Z is not weakly closed in Q but $m(H^*,\tilde{Q}) \le 2$. Then G is described in 19.2. If the pair G,Q is described in (2) then the conclusion of (2) can be seen to hold from 19.6.

If $p = 2$ and G is Co_1 or Sz then H^* contains the commutator group of the orthogonal group on $\tilde{Q}$. Hence H is transitive on singular points of $\tilde{Q}$ while $\tilde{U} = C_{\tilde{Q}}(T)$ is a singular point and $N_H(U)$ is transitive on $O_2(C_{H^*}(\tilde{U}))$. The first remark and the fact that Z is not weakly closed in Q show $U = ZZ^g$ for some $g \in G$. The second says $(Q^g \cap H)^* = O_2(C_{H^*}(\tilde{U}))$ and (1) holds as in the proof of 19.7.

Suppose $G \cong Aut(M(22))$. Then $Z(\tilde{T}) = U = \langle\tilde{d},\tilde{t}\rangle$ where $z = ad$ with a and d uniquely determined 3-transpositions, and $t = z^g$. In particuar d and $dt = da^gd^g$ are the product of one and three commuting 3-tranpositions, respectively and hence not conjugate to z. Therefore $z^G \cap U = \{z,t,tz\}$ and it remains to show $\mathcal{Y}(T) \subseteq U$ to verify that (6) holds. But from the proof of 19.7, $m(P^*) \ge 4$, where $P = Q^g \cap H$. Thus as $O_2(C_{H^*}(\tilde{t}))$ contains a unique subgroup $P_0^* \cong E_{16}$ and as P_0^* is weakly closed in T^*, we conclude $P^* = P_0^*$ and then $\mathcal{Y}(T) \subseteq U$ since $\tilde{U} = C_{\tilde{Q}}(P_0)$.

Suppose G,Q is as in (4). Then $\tilde{Q}$ is the sum of two dual 3-dimensional submodules $\tilde{Q}_i$ for $H^* \cong L_3(2)$. Therefore $\tilde{U} = Z(\tilde{T}) \cong E_4$ with $\tilde{U} = \langle\tilde{u}_1,\tilde{u}_2\rangle$ and $\tilde{u}_i \in \tilde{Q}_i$. Notice there is $X \in \mathfrak{X}(T) - z^G$ described in 19.4 and $C_G(ZX)$ is a 2-group. Therefore $\tilde{X} = \langle\tilde{u}_1\tilde{u}_2\rangle$ and $u_i \in z^G$. As is 19.7, $(Q^g \cap H)^* = P^* \cong E_2k$, $k \ge 2$, so as $H^* \cong L_3(2)$ it follows that if $z^g \in \mathcal{Y}(T)$ then P^* is weakly closed in T^*, completing the proof that (4) holds.

It remains only to consider certain groups in 19.2.3 and 19.2.5, which we find satisfy (1).

(19.9) Let G be He and $p = 2$. Then

(1) There exists a Sylow 2-group T of G containing Q and overgroups G_i of T in G, i = 1,2, with $G_i \cong S_6/Z_3/E_{64}$ and $G_i = N_G(A_i)$, where $A_i = O_2(G_i)$.

(2) $\mathcal{A}(T) = \{A_1, A_2\}$.

(3) A_1 is fused to A_2 in Aut(G).

(4) $Z(J(T)) = V = A_1 \cap A_2 \cong E_{16}$, $N_G(V) = G_1 \cap G_2$, and $\mathcal{S}(N_G(V))$ is the Shult space of $N_G(V)/A_1A_2 \cong \Omega_4^-(2)$.

Proof. Parts (1) and (3) follow easily from 4.1 in [14] for example. Part (2) follows from (1). Part (4) is a consequence of (1), (2), and 17.6.

(19.10) Let G be Co_3 and $p = 3$. Then $|T| = 3^7$, $V = J(T) \cong E_{3^5}$, and $N_G(V)/V \cong M_{11} \times Z_2$.

Proof. From 19.4 there is $E_{3^5} \cong A \leq T$ and $X \leq A$ with $A \trianglelefteq N_G(X)$ and $N_G(X)/A \cong Z_2 \times S_5$. As $H^* \cong Z_2/(Z_4 * SL_2(9))$, it follows that $m(A \cap Q) = 3$ and $AQ = T$. Then $(A \cap Q)/Z = C_{\tilde{Q}}(T)$ so $A \cap Q$ and $A = C_G(A \cap Q)$ are uniquely determined. Thus $A = V$.

As V is weakly closed in T, $M = N_G(V)$ controls fusion in V. Therefore $|M| = m_Z n_Z = m_X n_X$, where $m_U = |U^G \cap M|$ and $n_U = |N_M(U)|$. From section 5 in [23] there is only one more class Y^G of subgroups of G of order 3 and moreover $m_3(C_G(Y)) < 5$. Thus $m_Z + m_X = 121$ is the number of points of V. Next $m_X = m_Z n_Z / n_X = 6m_Z/5$, so $121 = m_Z + m_X = 11m_Z/5$. That is $m_Z = 55$, $m_X = 66$, and the order of $\bar{M}$ is that of M_{11}, where $\bar{M} = M/V\langle t \rangle$ and t is the involution in $N_M(X)$ inverting V.

Next $G \leq L \cong Co_1$. Let $T \leq S \in Syl_3(L)$. From the proof of 19.3, $V \cap Q = J \cap Q$, where $J = J(S)$, and $V = C_G(V \cap Q) \leq C_L(J \cap Q) = J$. Then $M \leq N_L(V) = C_L(V)(N_L(J) \cap N_L(V)) = (N_L(J) \cap N_L(V))J$. So $\bar{M} \leq N_L(J)/J\langle t \rangle \cong M_{12}$ by 19.3. Thus as the order of $\bar{M}$ is that of M_{11} it follows from the table of subgroups of M_{12} in [19] that $\bar{M} \cong M_{11}$.

20. The proof of Theorem B.

In this section Q is a large symplectic p-subgroup of the finite group G and Z = $\langle z \rangle$ = $\Omega_1(Z(Q))$. Assume L = $\langle z^G \rangle$ is nonabelian simple and G = LQ. Exclude case (2) of Theorem A. The major result of this section is:

Theorem B. Let $Q \leq M < G$ and let $K = \langle Q^M \rangle$. Then one of the following holds:

(1) $z \in O_p(M) = F^*(M)$.

(2) p = 2, $G \cong P\Omega_6^{\epsilon}(3)$, $K \cong P\Omega_5(3)$, and if $G \cong U_4(3)$ then M = K and |G:M| is even.

(3) p = 2, $G \cong G_2(3)$, and $M = K \cong G_2(2)$.

(4) p = 2, $G \cong P\Omega_8^+(3)$, and $M = K \cong \Omega_8^+(2)$.

(5) p = 2, $G \cong Co_2$, $K \cong U_6(2)$, and |G:M| is even.

(6) p = 2, $G \cong Co_1$, $K \cong U_6(2)$ or Co_2, and |G:M| is even.

(7) p = 2, $G \cong M(24)'$, $M = K \cong \Omega_{10}^-(2)$, and |G:M| is even.

(8) p = 2, $G \cong F_3$, $K \cong {}^3D_4(2)$, and |G:M| is even.

(9) p = 3, $G \cong F_2$ or M(23), and $K \cong P\Omega_8^+(3)$ or M(23).

(10) p = 3, $G \cong {}^2E_6(2)$ or M(22), and $K \cong P\Omega_7(3)$ or M(22).

(11) p = 3, $G \cong Co_1$, and $K \cong U_4(3), U_6(2), Mc, Co_2$, or Co_3.

(12) p = 3, $G \cong Co_2$, and $K \cong U_4(3), U_6(2)$, or Mc.

(13) p = 3, $G \cong Co_3$, and $K \cong U_4(3)$ or Mc.

(14) p = 3, $G \cong$ Mc or $U_6(2)$, and $K \cong U_4(3)$.

(15) p = 3, $G \cong F_4(2)$, and $K \cong L_4(3)$ or $Z_3/\Omega_8^+(2)$.

(16) p = 5, $G \cong F_2$, and $K \cong F_5$.

(17) p = 5, $G \cong$ Ly, and $K \cong G_2(5)$.

(18) p = 3, $G \cong A_{11}$ or A_{10}, and $K \cong A_9$ or A_{10}.

(19) p = 3, $G \cong {}^2F_4(2)'$ and $K \cong L_3(3)$.

(20) p = 3, $G \cong Sp_6(2)$, and $K \cong PSp_4(3)$ or $U_3(3)$.

(21) p = 3, $G \cong {}^3D_4(2)$, and $K \cong U_3(3)$.

(22) p = 5, $G \cong Co_1$, and $K \cong U_3(5)$, Mc, HS, Co_3, or Co_2.

(23) p = 5, $G \cong Co_2$ or Co_3, and $K \cong$ Mc or HS.

(24) p = 5, $G \cong Co_2, Co_3$, Mc, HS, or Ru and $K \cong U_3(5)$.

(25) $p = 7$, $G \cong ON$, and $K \cong L_3(7)$.

(26) $p = 7$, $G \cong M(24)'$, and $K \cong He$.

(27) $p = 11$, $G \cong J_4$, and $K \cong U_3(11)$.

The proof of Theorem B is relatively straightforward. By 17.4 we may assume that either $F^*(K)$ is a nonabelian simple group or $Q \cong D^2$ and $F^*(K) \cong A_5 \times A_5$. The possiblities for (G,Q) are enumerated in Theorem A. Observe first that we may assume $G \notin Chev(p)$ by 10.5. Thus if $F^*(K)$ is not simple then as $Q \cong D^2$ we have $G \cong M_{22}, M_{12}, G_2(3)$, or $P\Omega_6^\epsilon(3)$. This is impossible as 5^2 does not divide the order of G.

So $F^*(K)$ is simple and the possibilities for (K,Q) are also enumerated in Theorem A. Table B lists the possibilities for Q,G, and K. Denote by $\mathfrak{A}(Q)$ the collection of groups listed in the G-K row of Table B corresponding to Q. Thus for example $\mathfrak{A}(D^{11}) = \{L_{13}(2), F_2\}$. Define two partial orders on $\mathfrak{A}(Q)$. Define Y b X if the order of Y divides the order of X and define $Y \to X$ if there is an injection $\alpha : Y \to X$ which is the identity on Q. Thus $Y \to X$ implies Y b X. To establish Theorem B amounts to determining the pairs (K,L) with $K \to L$. Denote by $\mathcal{C}(X)$ the collection of groups Y with $Y \to X$ and by $\mathfrak{B}(X)$ the groups Y with Y b X. Hence $\mathcal{C}(X) \subseteq \mathfrak{B}(X)$.

Table 20.1 lists the groups $G \notin Chev(p)$ and those K in $\mathfrak{B}(G)$ which do not appear in Theorem B. It remains to eliminate these cases and to establish the extra remarks in Theorem B. Let $H = N_G(Z)$, $H^* = H/Q$, and $\tilde{H} = H/Z$.

If $G \cong M(24)'$ and $K \cong \Omega_{10}^-(2)$ then $K = M$ by 15.9.5 and hence $|G:M|$ is even.

Assume $Q \cong Q^4$. If $K \cong \Omega_8^+(2)$ or ${}^3D_4(2)$ there exists $g \in K - H$ with $z^g \in Q$ and $m(Q \cap Q^g) = 5$. But if $G \cong Co_2$ then Z is weakly closed in Q with respect to G, while if $G \cong Co_1$ then $Q \cap Q^g = \langle Z, Z^g\rangle$ for each $Z \neq Z^g \leq Q$. So if G is a Conway group then K is not $\Omega_8^+(2)$ or ${}^3D_4(2)$. Notice also that if G is a Conway group then $|G|_2 > |Aut(K)|_2$ for each $K \in \mathcal{C}(G) - \{G\}$, so Theorem B is established in these cases. If G is F_5 then $|H|_3 = 9$ while if $K \cong \Omega_8^+(2)$ then $|H \cap K|_3 = 27$, eliminating this inclusion. This leaves $G \cong F_3$. $|G|_2 > |Aut({}^3D_4(2))|_2$, so we may take K to be $\Omega_8^+(2)$ or $L_6(2)$. Here there is y of order 3 in K with $C_K(y) \cong Z_3 \times L_4^\epsilon(2)$, whereas no element of order 3 centralizes an $L_4^\epsilon(2)$-section in F_3.

Suppose $Q \cong Q^3$, $G \cong Sz$, and $K \cong U_5(2)$. Then there is y of order 5 in K centralizing a $SU_3(2)$-subgroup of K, so $|C_K(y)|_3 \geq 27$. This is impossible as $|C_G(y)|_3 \leq 9$ for each y of order 5 in $G \cong Sz$.

Table B

Q	G or K
D^2	$G_2(2)$, $L_4^\epsilon(2)$, A_9, A_{10}, $G_2(3)$, $P\Omega_6^\epsilon(3)$, M_{22}
QD	J_2, J_3
Z_4*D^2	HS, $Z_2/U_4(3)$
Q^3	$U_5(2)$, Sz
D^3	$L_5(2)$, $E_4/U_4(3)$, M_{24}, He, Aut(HS)
D^4	$L_6^\epsilon(2)$, $\Omega_8^\epsilon(2)$, ${}^3D_4(2)$, $P\Omega_8^+(3)$, Co_1, Co_2, F_3, F_5
D^5	Aut(M(22)), $L_7(2)$
D^6	$L_8^\epsilon(2)$, $\Omega_{10}^\epsilon(2)$, J_4, M(24)′
D^{11}	$L_{13}(2)$, F_2
D^{12}	$L_{14}^\epsilon(2)$, $\Omega_{16}^\epsilon(2)$, F_1
3^{1+12}	F_1, $L_8^\epsilon(3)$, $P\Omega_{10}^\epsilon(3)$, $PSp_{14}(3)$
3^{1+10}	M(24)′, $L_7^\epsilon(3)$, $PSp_{12}(3)$, $P\Omega_9(3)$
3^{1+8}	M(23), F_2, $L_6^\epsilon(3)$, $P\Omega_8^\epsilon(3)$, ${}^3D_4(3)$, $PSp_{10}(3)$
3^{1+6}	${}^2E_6(2)$, M(22), $L_5^\epsilon(p)$, $PSp_8(p)$, $P\Omega_7(p)$
5^{1+6}	F_1, $L_5^\epsilon(p)$, $PSp_8(p)$, $P\Omega_7(p)$
3^{1+4}	$Z_3/\Omega_8^+(3)$, $U_6(2)$, $F_4(2)$, Co_1, Co_2, Co_3, Mc, F_5, $PSp_6(p)$, $L_4^\epsilon(p)$, $G_2(p)$
5^{1+4}	Ly, F_5, F_2, $PSp_6(p)$, $L_4^\epsilon(p)$, $G_2(p)$
7^{1+4}	F_1, $PSp_6(p)$, $L_4^\epsilon(p)$, $G_2(p)$
3^{1+2}	A_9, A_{10}, A_{11}, $PGL_3(4)$, $U_3(8)$, ${}^3D_4(2)$, ${}^2F_4(2)'$, $Sp_6(2)$, M_{12}, $L_3^\epsilon(p)$, $PSp_4(p)$
5^{1+2}	$U_5(4)$, Co_1, Co_2, Co_3, HS, Mc, Ru, F_3, $L_3^\epsilon(p)$, $PSp_4(p)$

Q	G or K
7^{1+2}	$M(24)'$, He, ON, $L_3^\epsilon(p)$, $PSp_4(p)$
11^{1+2}	J_4, $L_3^\epsilon(p)$, $PSp_4(p)$
13^{1+2}	F_1, $L_3^\epsilon(p)$, $PSp_4(p)$

Table 20.1

Q	G	K
D^4	Co_1, Co_2, F_3, F_5	$\Omega_8^+(2)$
D^4	F_3	$L_6(2)$
D^4	Co_1	${}^3D_4(2)$
Q^3	Sz	$U_5(2)$
D^2	*	*
3^{1+12}	F_1	$P\Omega_{10}^+(3)$
5^{1+6}	F_1	$P\Omega_7(5)$
3^{1+4}	Co_1	$PSp_6(3)$, $L_4(3)$
3^{1+4}	Co_1, Co_2, F_5	$Z_3/\Omega_8^+(2)$
3^{1+4}	F_5	Mc
3^{1+4}	$Z_3/\Omega_8^+(3)$, F_5, $F_4(2)$	$U_4(3)$
5^{1+4}	F_2	$U_4(5)$
7^{1+4}	F_1	$L_4(7)$, $G_2(7)$
3^{1+2}	A_n, $Sp_6(2)$	$PGL_3(4)$
3^{1+2}	A_n	$PSp_4(3)$
3^{1+2}	A_{11}	M_{12}
3^{1+2}	A_n, $PGL_3(4)$	$U_3(3)$
3^{1+2}	${}^3D_4(2)$	$L_3(3)$
3^{1+2}	$Sp_6(2)$	A_9
5^{1+2}	Co_1	$PSp_4(5)$
5^{1+2}	F_3	$L_3^\epsilon(5)$

Assume $Q \cong Q^2$. If G is $G_2(3)$ or $P\Omega_6^{\epsilon}(3)$ we appeal to 9.8. So if $Q \cong Q^2$ then we are left with the case $G \cong M_{22}$. Then as $K \in \mathcal{B}(G)$, $K \cong L_4(2)$. But $L_4(2)$ has an element of order 15 while M_{22} does not. This completes the discusion of the case p = 2.

Suppose p = 3, G is F_1, and $K \cong P\Omega_{10}^{+}(3)$. Let $H_K = H \cap K$. Then $O^{3'}(H^*) \cong Sz/Z_2$ and $O^{3'}(H_K^*) \cong SL_2(3)*\Omega_6^{+}(3)$. Let $h^* \in H_K^*$ be of order 4 in the $SL_2(3)$ factor. Then its image in Sz is an involution lifting to an element of Sz/Z_2 of order 4. Hence the class of this image is determined and its centralizer in Sz is $Z_2/(L_3(4) \times E_4)$. As this centralizer has no $\Omega_6^{+}(3)$ section, we have a contradiction.

Similarly if p = 5, $G \cong F_1$, and $K \cong P\Omega_7(5)$, then $O^{5'}(H^*) \cong J_2/Z_2$ and $O^{5'}(H_K^*) \cong SL_2(3)*\Omega_3(5)$. Hence if $h^* \in H_K^*$ is an involution in the $\Omega_3(5)$ factor then the centralizer of its image in J_2 is A_5/QD, and in particular contains no $E_4 \times A_5$ subgroup as does the centralizer in $O^{5'}(H_K^*)$. So this case is eliminated.

Assume $Q \cong 3^{1+4}$. Suppose G is a Conway group. We saw in 19.5 and 19.6 that Z is weakly closed in Q with respect to G, but not with respect to $L_4(3)$ or $Z_3/\Omega_8^{+}(2)$. So K is neither of these groups. Therefore G is Co_1 and K is $PSp_6(3)$. Let $T \in Syl_3(K)$. Then $J(T) \cong E_{3^6}$ and $N_K(J(T)/J(T) \cong GL_3(3)$. But by 19.3, $N_G(J(T))/J(T) \cong M_{12}/Z_2$, while $L_3(3)$ is not a subgroup of M_{12}.

Let $G \cong F_5$ and $K \cong Z_3/\Omega_8^{+}(2)$, and r a 2-central involution in K. Then 3^4 divides the order of $C_K(r)$ but does not divide the order of the centralizer of an ivolution of G, a contradiction.

Notice that 19.6 immediately eliminates the containment of $U_4(3)$ or Mc in $Z_3/\Omega_8^{+}(3)$, F_5, and $F_4(2)$, and thus completes the analysis of the case $Q \cong 3^{1+4}$.

F_2 is of 5-rank 3 while $U_4(5)$ is of 5-rank 4, so $U_4(5)$ is not contained in F_2.

Suppose p = 7, $G \cong F_1$, and $K \cong G_2(7)$ or $L_4(7)$. Then there is X of order 3 in K such that $Y = E(C_K(X)) \cong SL_3(7)$. Then 7^3 divides $|C_G(X)|$ so $C_G(X) \cong M(24)'/Z_3$. But $L_3(7)$ is not a subgroup of $M(24)'$ by an order argument.

$PGL_3(4)$ has no faithful permutation representation of degree 11 or linear representation of degree 6 over GF(2), and hence is not contained in A_{11} or $Sp_6(2)$. Similarly $PSp_4(3)$, M_{12}, and $U_3(3)$ are not in A_{11}, $U_3(3)$ is not in $PGL_3(4)$, and A_9 is not in $Sp_6(2)$.

If p = 3 and $G \cong {}^3D_4(2)$ then Z is weakly closed in Q. This is not true in $L_3(3)$, eliminating this containment.

We are left with the cases $Q \cong 5^{1+2}$ and $(G,K) = (Co_1, PSp_4(5))$ or $(F_3, L_3^{\epsilon}(5))$.

Suppose G is Co_1 and K is $PSp_4(5)$. Then $H^* \cong GL_2(5)$ and if $Q \leq T \in Syl_5(K)$ then V $= J(T) \cong E_{5^3}$ with $N_K(V)$ irreducible on V. So by 17.6, $N_G(V)/V \cong Z_4 \times PO_3(5)$. Now $O^2(N_K(V)) = O^2(N_G(V))$ and $O^2(N_K(Q)) = O^2(H)$ with $H = N_H(V)O^2(H)$, so H acts on K = $\langle O^2(N_K(V)), O^2(N_K(Q))\rangle$. This forces HK to be $GSp_4(5)$. In particular there is an involution $s \in HK - K$ with $E(C_K(s)) \cong L_2(25)$. Therefore 13.25 divides the order of $C_G(s)$, so $E(C_G(s)) \cong G_2(4)$. This says $L_2(25)$ is a subgroup of $Sp_6(4)$ which is easily seen not to be the case from the structure of a Sylow 5 normalizer in $L_2(25)$.

So take $G \cong F_3$ and $K \cong L_3^{\epsilon}(5)$. Then $H^* \cong Z_2/(Z_4{*}SL_2(3))$.

Assume $\epsilon = +1$ and let V_1 and V_2 be the radicals of the maximal parabolics of K containing Q. Then $H_K^* \cong Z_4^2$. Thus a Sylow 2-subgroup S of H_K is a uniquely determined subgroup of index 2 in a Sylow 2-subgroup R of H, and from the structure of H, V_1 and V_2 are the unique subgroups of index 5 in Q invariant under S. Moreover $(V_1)^r = V_2$ for $r \in R - S$ and $O^2(N_K(V_i)) = O^2(N_G(V_i))$, so S acts on K = $\langle O^2(N_K(V_i)) : i = 1,2\rangle$. Therefore $SK \cong PGL_3(5)$ so there is an involution $a \in KS - K$ with $E(C_K(a)) \cong L_2(5)$. We now show $C_G(a)$ contains no subgroup $X \cong L_2(5)$, to derive a contradiction. For let $A = C_G(a)$, $E = O_2(A)$, and $\bar{A} = A/E$. $\bar{A} \cong A_9$ so $\bar{X}$ fixes 3 or 4 points in the representation of $\bar{A}$ on 9 letters. In either case if x is an involution in X then $\bar{x}$ fixes 5 points. To complete the proof we observe that by 3.11 in [32], the coset xE contains no involutions.

Finally consider the case $\epsilon = -1$. Let a be the involution in H_K and as above let $A = C_G(a)$, $E = O_2(A)$, and $\bar{A} = A/E$. $E(C_K(a)) = J \cong SL_2(5)$ with $C_K(a)$ the extension of J by an involution b inducing an outer automorphism on J. As above, $\bar{J} \cong A_5$ fixes 3 or 4 points in the representation of $\bar{A}$ on 9 letters. In the latter case each involution $\bar{b} \in \bar{A}$ inducing an outer automorphism on $\bar{J}$ fixes 5 points. But by an earlier remark, bE contains no involutions, whereas b is one such.

So $\bar{J}$ fixes 3 points. Hence $\tilde{E} = E/\langle a\rangle$ is the sum of two $\Omega_4^-(2)$ modules for $\bar{J}$, (An A_8-subgroup of $\bar{A}$ has two 4-dimensional chief factors on $\tilde{E}$.) so that $H^1(\bar{J},\tilde{E}) = 0$. Therefore $N_{\bar{A}}(\bar{J}) = N_A(J)E/E \cong Z_2/(Z_3 \times A_5)$. In particular a subgroup X of order 3 in $C_A(J)$ acts on $\langle J,Q\rangle = K$. Thus $KX \cong PGU_3(5)$. But then $[b,X] = 1$, whereas $\bar{b}$ inverts $\bar{X}$ from the structure of $N_{\bar{A}}(\bar{J})$.

The proof of Theorem B is at last complete.

(20.1) If $Q \leq M \leq G$ and Z is the unique elementary abelian maximal $(H \cap M)$-invariant subgroup of Q, then either

(1) $M \leq H$, or

(2) $O_p(M) = 1$.

Proof. If $O_p(M) \neq 1$ then by 17.4, $L = \langle Z^M \rangle$ is abelian. Then by hypothesis $L \cap Q = Z$, so by 17.6.5, $L = Z$. That is $M \leq H$.

(20.2) Assume Q is of width at least 2, let $H = N_G(Z)$, and let $M \in \mathcal{N}(H)$. Then one of the following holds:

(1) $H = M$.

(2) $G \cong L_n(p)$ and M is the parabolic stabilizing the point or hyperplane fixed by H in the projective space for G.

(3) $Q \cong D^3$, $G \cong M_{24}$, and $M \cong A_8/E_{16}$.

(4) $Q \cong D^2$, $G \cong M_{22}$, and $M \cong A_6/E_{16}$.

(5) $Q \cong D^2$, $G \cong P\Omega_6^+(3)$, and $M \cong P\Omega_5(3)$.

Proof. Let $K = \langle Q^M \rangle$ and $X = F^*(M)$. Assume $H \neq M$. By 17.4 and Theorem B either X is a nonabelian simple group and $K = XQ$ or $X = O_p(M)$ and $V = \langle Z^M \rangle$ is abelian.

Assume the former. Then K is described in Theorem B and $H = N_M(Z)$. Comparing the order of H and $N_M(Z)$ for the various possiblities for K and keeping in mind that $M \leq \mathrm{Aut}(X)$, we conclude that either (5) holds or $Q \cong 3^{1+8}$, $G \cong M(23)$, and $X \cong P\Omega_8^+(3)$. In the latter case an E_{27}-subgroup of $(H \cap M)/O_{3,2}((H \cap M)$ is normal in that group, whereas $H/O_{3,2}(H)$ has no such normal subgroup.

So we may assume $X = O_p(M)$ and $V \neq Z$ is abelian. If $F^*(G) \in \mathrm{Chev}(p)$, then by 10.5, M is a parabolic of G. Then as $H < M$, we conclude (2) holds. So assume $F^*(G) \notin \mathrm{Chev}(p)$. By 19.1 and 20.1, one of the following holds: (If G is $G_2(3)$ or HS then Z is the unique elementary abelian normal subgroup of H, even though H is not irreducible on $\tilde{Q}$.)

(a) $Q \cong D^2$, $G \cong M_{22}$, $m(V \cap Q) = 3$, and $m(V) \leq 4$.

(b) $Q \cong D^2$, $G \cong M_{12}$, and $E_8 \cong V \leq Q$.

(c) $Q \cong D^5$, $G \cong \mathrm{Aut}(M(22))$, and $E_4 \cong V \leq Q$.

(d) $Q \cong D^3$, $G \cong M_{24}$ or He, and $E_{16} \cong V \leq Q$.

(e) $Q \cong 3^{1+4}$, $G \cong Z_3/\Omega_8^+(2)$, and $E_9 \cong V \leq Q$.

In cases (c) and (e), Z is weakly closed in the unique normal E_{p^2}-subgroup V of H, contradicting $V = \langle Z^M \rangle$. If G is a Mathieu group appeal to the tables in [19]. Finally if G is He then 17.6 says $N_G(V)/V = GL(V)$. In particular $N_G(V)$ contains a Sylow 2-group T of G. By 19.9, T has just two elementary abelian subgroups A_i of maximal rank. Further $m(A_i) = 6$, $U = A_1 \cap A_2$ is of rank 4, and $N_G(U)/A_1A_2 \cong \Omega_4^+(2)$. This is incompatible with the structure of $N_G(V)$. For example $Z_2(T) \leq U$ so $J(T) \leq C_G(Z_2(T) \cap V) = O_2(P)$, where P/V is the parabolic of $N_G(V)/V$ isomorphic to $\Omega_4^+(2)/E_{16}$. But then $J(T) = O_2(P)$ by an order argument, whereas $m(Z(O_2(P))) = 2$.

(20.3) Assume Q is of width at least 2 and let $H = N_G(Z)$ and $T \in Syl_p(H)$. Then one of the following holds:

(1) $\mathcal{Y}(T)$ is nonempty and $G = \langle H, Q^g \rangle$ for each $Z^g \in \mathcal{Y}(T)$.

(2) $G \cong L_n(p)$, M_{24}, M_{22}, or $P\Omega_6^+(3)$, with $p = 2$ in the last three cases, and $G = \langle H, Q^g : Z^g \in \mathcal{Y}(T) \rangle$.

(3) $G \cong L_4(3)$ or Co_2, $p = 2$, and $\mathcal{Y}(T)$ is empty.

(4) $Q \cong 3^{1+4}$, $G \cong Co_1, Co_2, Co_3, Mc$, or $U_6(2)$, and $\mathcal{Y}(T)$ is empty.

Proof. If H is maximal in G and $Z^g \in \mathcal{Y}(T)$ then certainly $G = \langle H, Q^g \rangle$. Further if $\mathcal{Y}(T)$ is empty then 19.8 completes the proof.

So assume $H < M \in \mathcal{N}(T)$. Then G and M are described in 20.2. We find $M = \langle H, Q^g \rangle$ for suitable $Z^g \in \mathcal{Y}(T)$. Further there is $K \in \mathcal{N}(T) - \{M\}$ with $K = \langle H \cap K, Q^h \rangle$ for suitable $Z^h \in \mathcal{Y}(T)$. Therefore (2) holds.

21. Large extraspecial subgroups of width 1.

In this section p is an odd prime, Q is a large extraspecial p-subgroup of the finite group G, and $Z = \langle z \rangle = Z(Q)$. Assume $L = \langle z^G \rangle$ is a nonabelian simple group and $Q \cong p^{1+2}$. Recall that G is described in Theorem A.

(21.1) Either

(1) Q is the unique normal p^{1+2}-subgroup of $N_G(Z)$, or

(2) $G \cong U_3(8)$ or $U_5(4)$ and there are exactly p normal p^{1+2}-subgroups of $N_G(Z)$. These subgroups are fused in Aut(G).

Proof. Let $Q \le T \in Syl_p(G)$. Usually $Q = T$ so that (1) holds. If not G is A_n, $U_3(8)$, ${}^3D_4(2)$, $Sp_6(2)$, $U_5(4)$, or Co_1. If G is A_n or $Sp_6(2)$ then $T \cong Z_3wrZ_3$, so Q is the unique 3^{1+2}-subgroup of T and certainly (1) holds. If G is ${}^3D_4(2)$ or Co_1 then $O_p(N_G(Z))$ is of order p^3 so that (1) holds. Finally if G is $U_3(8)$ or $U_5(4)$ then T has a unique normal E_{p^2}-subgroup X and $N_{Aut(G)}(T)/X$ has Sylow p-subgroup p^{1+2} with center J(T)/X, so (2) holds.

(21.2) Let Δ be the set of subgroups X of Q of index p generated by conjugates of Z. Then Table 21.2 gives the order of Δ and the number of orbits of $N_G(Q)$ on Δ. If there is more than one orbit, then the orbits are fused in $N_{Aut(G)}(Q)$.

Proof. This can be determined from the list of normalizers of subgroups of order p in section 5 of [23]. (And from the corresponding information when G is not sporadic.)

(21.3) If $X \in \Delta$ then $O^{p'}(N_G(X)) \cong SL_2(p)/E_{p^2}$ and $C_G(X) = X$.

Proof. Let $X = ZZ^g$ and $Y = \langle Q,Q^g \rangle$. Then $Y \cong SL_2(p)/E_{p^2}$ by 17.8. Notice $Aut_Y(X) = O^{p'}(Aut(X))$. Therefore $O^{p'}(N_G(X)) = YC_G(X)$. Also as $Y \trianglelefteq N_G(X)$, $YC_G(X) = Y \times C_G(Y)$.

Table 21.2

p	G	$\|\Delta\|$	n
3	A_n	3	1
3	$U_3(8)$	3	1
3	${}^3D_4(2)$	0	–
3	${}^2F_4(2)'$	4	2
3	$Sp_6(2)$	0	–
3	M_{12}	2	2
5	$U_5(4)$	5	1
5	Co_1	0	–
5	Co_2	0	–
5	Co_3	0	–
5	HS	0	–
5	Mc	0	–
5	Ru	2	1
5	F_3	6	1
7	$M(24)'$	6	2
7	He	3	1
7	ON	4	2
11	J_4	0	–
13	F_1	6	1

Finally as $Z = C_G(Q)$, $C_G(Y) = 1$, so the lemma is established.

Theorem 21.4. Table 21.4 lists those $M \in \mathcal{N}(Q)$ with $F^*(M) \neq O_p(M)$. Column 1 lists p, column 2 lists G, column 3 lists the isomorphism type of M, column 4 lists the number of conjugates of M containing a Sylow p-subgroup of M, and column 5 lists the number of orbits of G on subgroups isomorphic to M. If there are two orbits, these orbits are fused in Aut(G). Finally Q is Sylow in M unless G is A_n or G is $Sp_6(2)$ and $M \cong O_6^-(2)$.

Corollary 21.5. Table 21.5 lists those $X \in \Delta$ with $N_G(X) \in \mathcal{N}(Q)$. If such an X exists then $N_G(Q)$ is transitive on Δ, unless G is M_{12} where $N_{Aut(G)}(Q)$ is transitive on Δ.

Corollary 21.6. Table 21.6 lists those $M \in \mathcal{N}(Q)$ which do not appear in 21.4 or 21.5. T is a Sylow p-subgroup of G containing Q. $M = N_G(J(T))$ only when $J(T) \cong E_{p^3}$, in which case M is described in 17.6.6.(ii).

Notice that 21.2 and 21.4 imply 21.5 and 21.6, with 17.6 also used to prove 21.6. Theorem 21.4 is established in the rest of this section and in the next section.

We begin the proof of Theorem 21.4. Assume that $LQ \leq G$ with $L = F^*(LQ) \neq O_p(LQ)$. Then G and LQ are described in Theorem B. It remains to determine when there exists a subgroup LQ with $N_G(L)$ maximal, to determine $N_G(L)$, and to determine the number of orbits of G on subgroups isomorphic to LQ. Notice that if $Q \leq S \in Syl_p(N_G(L))$ then S is in $|N_G(S):N_G(S) \cap N_G(L)|$ conjugates of $N_G(L)$, allowing us to fill in the fourth column of Table 21.4. We will see during the proof that if there are two orbits, then these orbits are fused in Aut(G).

If G is A_n it is easy to fill in Table 21.4 using the representation of G on n letters.

Suppose G is $Sp_6(2)$. Then $L = LQ \cong PSp_4(3)$ or $U_3(3)$ by Theorem B. G is the isometry group of (V,f) where V is a 6-dimensional space over GF(2) and f is a nondegenerate symmetric bilinear form on V. Up to equivalence there is a unique faithful representation of $PSp_4(3)$ and $U_3(3)$ on V, each preserves a conjugate of f, and each is absolutely irreducible. Therefore by 2.5 and 10.5 in [3], G is transitive on its subgroups isomorphic to L, such subgroups exist, and $N_G(L) \cong Aut(L) \cong$

Table 21.4

p	G	M	Conjugates over S	Orbits on M-subgroups
3	A_{10}	A_9	1	1
3	A_{11}	A_{10}	2	1
		S_9	1	1
3	$^3D_4(2)$	$G_2(2)$	3	1
3	$^2F_4(2)'$	$Aut(L_3(3))$	2	2
3	$Sp_6(2)$	$O_6^-(2)$	1	1
		$G_2(2)$	3	1
5	Co_1	Co_2	5	1
		Co_3	10	1
5	Co_2	Mc	4	1
		Aut(HS)	3	1
5	Co_3	Aut(Mc)	1	1
		$Aut(U_3(5))$	1	1
		HS	3	1
5	Mc	$U_3(5)$	3	1
5	HS	$Z_2/U_3(5)$	2	2
5	Ru	$Z_2/U_3(5)$	2	1
7	ON	$Z_2/L_3(7)$	2	2
7	M(24)′	Aut(He)	2	2
11	J_4	$Z_2/U_3(11)$	3	1

Table 21.5

p	G	$\|\Delta\|$
3	A_9	3
3	$U_3(8)$	3
3	M_{12}	2
5	$U_5(4)$	5
5	Ru	2
5	F_3	6
7	He	3
13	F_1	6

Table 21.6

p	G	M
3	A_9	$N_G(J(T))$
3	$U_3(8)$	$N_G(Q)$
3	${}^3D_4(2)$	$N_G(Q)$
5	$U_5(4)$	$N_G(J(T))$
5	Co_1	$N_G(Q)$, $N_G(J(T))$
5	Co_3	$N_G(Q)$, $N_G(J(T))$
5	Co_2	$N_G(Q)$
5	Mc	$N_G(Q)$
5	Ru	$N_G(Q)$
5	F_3	$N_G(Q)$
7	He	$N_G(Q)$
11	J_4	$N_G(Q)$
13	F_1	$N_G(Q)$

$O_6^-(2)$ or $G_2(2)$. Thus Theorem 21.4 holds in this case. Notice Theorem B insures the maximality of $N_G(L)$; this comment will not be made again.

On page 396 of [34], Seitz observes that if G is the Tits group then there exists $X \le G$ with $X \cong Aut(L_3(3))$. We may take $Q \le X$. As $|N_X(Q)| = |N_G(Q)|$, $N_G(Q) \le X$. As $N_G(Q)$ has two orbits Δ_1 and Δ_2 on Δ and these orbits are fused in Aut(G), we have (up to choice of notation) $X = \langle N_G(D), N_G(Q) : D \in \Delta_1 \rangle$, so that there are two $Aut(L_3(3))$-subgroups of G containing Q, and these subgroups are fused in Aut(G) but not in G. So Theorem 21.4 holds in this case.

Let G be ${}^3D_4(2)$, so that $L = LQ \cong U_3(3)$. Certainly one such subgroup exists as the centralizer of some outer automorphism of G is $G_2(2)$. Let t be the involution inverting Q/Z and $R = O_2(C_G(t))$. Then $C_L(t) \cong Z_2/(Z_4 * SL_2(3))$ and $C_G(t) \cong L_2(8)/D^4$. Thus $\mathcal{U}^*_{C_G(t)}(Z,2) = \{R\}$, so $O_2(C_L(t)) \le R$. $R_0 = C_R(Z) \cong Q_8$ and R_0 centralizes [R,Z], so R_0 acts on $\langle Q,[R,Z]\rangle = L$. Hence $X = N_G(L) = R_0L \cong G_2(2)$.

Let s be a non-2-central involution in $N_X(Z)$. Then $s \in X - L$ so $C_L(s) \cong S_4$ and if $r \in O_2(C_L(s))^\#$ then $r \in t^L$ with $s \in O_2(C_X(r)) \le O_2(C_G(r))$. Also s centralizes $x \in Q - Z$ so as x is not 3-central, s is not 2-central. It turns out that as s is in the non-2-central class, $\Gamma = \{t^G : s \in R^g\}$ is of order 3. Hence $\Gamma = O_2(C_L(s))^\#$, so $L = \langle L,\Gamma\rangle$ is uniquely determined by s and Q. Finally $N_X(Z) = QR_0\langle s\rangle$ with $R_0\langle s\rangle \cong SD_{16}$ a Sylow 2-group of $N_G(Z)$. Thus s is determined up to conjugacy in $N_G(Z)$ so G is transitive on its $G_2(2)$-subgroups, and Theorem 21.4 holds in this case.

This completes the analysis when p = 3. The case p = 5 is the most complex and is postponed till the next section.

Let p = 7, so that G is ON or M(24)′. Assume first that G is ON. Then in section 10 of [31], O'Nan uses the Brauer trick to show that if $D \in \Delta$ then $\langle N_G(Q), N_G(D)\rangle = X$ is a proper subgroup of G. Therefore by Theorem B, $X \cong Z_2/L_3(7)$. It follows from 21.2 that Theorem 21.4 holds in this case.

So take G to be M(24)′. From 15.16.4 there exists $L \le G$ with $L \cong He$. The proof that there are exactly two classes of such subgroups was postponed until now. Assume L is such a subgroup. Without loss $Q \le L$. By 21.2, the set Γ of subgroups X of Q of index 7 with $X^\# \subseteq z^L$ is of order 3 and $N_L(Q)$ is transitive on Γ. Again by 21.2, Γ is $N_G(Q)$-invariant and $\Delta = \Gamma \cup \Gamma^\alpha$ for some $\alpha \in N_{Aut(G)}(Q)$. Now $L = \langle O^{p'}(N_G(D)) : D \in \Gamma\rangle$ by Theorem B, so $N_G(Q) \le N_G(L)$ and hence $N_G(L) \cong Aut(He)$. Further we see L and L^α are the unique He-subgroups of G containing Q, and thus that Theorem 21.4 holds.

This leaves the case $p = 11$, $G \cong J_4$, and $L \cong U_3(11)$. I'll prove there is at most one class of $U_3(11)$ subgroups. The proof of the existence of such subgroups is left to 33.4.7. So assume $U_3(11) \cong L \leq G \cong J_4$, let t be an involution inverting Q/Z, $H = C_G(t)$, $H^* = H/O_2(H)$, and $\tilde{H} = H/\langle t\rangle$. Then $K = E(C_L(t)) \cong SL_2(11)$ and $V = C_L(K) \cong Z_4$. By 33.4.1, t is 2-central and $H \cong Aut(M_{22})/Z_3/D^6$, while by 33.4.2, $C_H(V)^* \cong PGL_2(11)$. Therefore $C_H(V)^* = C_L(V)^*$

Let $X \in Syl_3(O_{2,3}(H))$, so that X is of order 3 and $[X^*,K^*] = 1$. Let $U = \langle V^X\rangle$. $[\tilde{V},K^*] = 0$ so $[\tilde{U},K^*] = 0$. Hence U centralizes $K = K^\infty$. Of course $U \cong Q_8$ and U acts on $\langle K,Q\rangle = L$, so $UL \cong Z_2/U_3(11)$. Similarly $C_G(K) \leq N_G(L)$ so $C_G(K) \leq C_{Aut(L)}(K) \cong Z_2/Z_{12}$. On the other hand $C_G(K)^* \leq C_{H^*}(K^*) = X^*$, so $C_G(K) \leq XU \cong SL_2(3)$, and hence $U = C_G(K)$ and $UL = N_G(L)$.

Let $W = C_{O_2(H)}(U)$. Then $C_G(U) = KW = C_{KW}(X)W$. So as $[X,K] \neq 1$, there is more than one class of complements to $\tilde{W}$ in $\tilde{K}\tilde{W}$. Claim $H^1(K^*,\tilde{W})$ is of dimension 1 over $GF(4)$. For there is an indecomposable $GF(4)K^*$-module I with $[K^*,I] = \tilde{W}$ and $H^1(K^*,\tilde{W}) \cong I/\tilde{W}$. (cf section 17 in [6]) Let k be of order 5 in $N_K(Q)$ and $s \in K$ invert k. $\tilde{W} = [\tilde{W},Z]$ so $\dim(C_{\tilde{W}}(k)) = 1$, $I = \tilde{W}\oplus C_I(Z)$, and $C_I(k) = C_{\tilde{W}}(k)\oplus C_I(Z)$ with $[C_I(k),s] \leq C_{\tilde{W}}(k)$. So if $\dim(I/\tilde{W}) > 1$ then $0 \neq C_I(Z)\cap C_I(s) = C_I(K) = 0$, a contradiction.

We conclude from the claim that there are four classes of complements to $\tilde{W}$ in $\tilde{K}\tilde{W}$, X fixes one class, and X is transitive on the remaining classes. Hence K is determined up to conjugacy under $N_G(Q)$, so the conjugacy class of $L = \langle K,Q\rangle$ is determined.

The proof of Theorem 21.4 has been reduced to the case $p = 5$. That case is dealt with in the next section.

22. Large extraspecial 5-subgroups of width 1.

In this section assume the hypothesis of section 21 with $p = 5$. Also let t be an involution inverting Q/Z, $H = C_G(t)$, $H^* = H/O_2(H)$, and $\tilde{H} = H/\langle t\rangle$.

(22.1) Let $G \cong Ru$. Then

(1) G contains exactly one conjugacy class of subgroups $K \cong SL_2(5)$ with $Aut_G(K) \cong PGL_2(5)$.

(2) G contains exactly one conjugacy class of subgroups isomorphic to $U_3(5)$. If L is such a subgroup then $|N_G(L):L| = 2$.

Proof. Let Ω be the set of subgroups of G isomorphic to $U_3(5)$ and Γ the set of subgroups described in (1). By Conway [18], there is $L \in \Omega$ such that there is $s \in G$ inducing an outer automorphism on L. We may take $Q\langle t\rangle \leq L$. So Ω is nonempty. Let $L_1 \in \Omega$. Then L_1 contains a subgroup $K_1 \cong SL_2(5)$ and there is $u_1 \in L_1$ inducing an outer automorphism on K_1. Thus $K_1 \in \Gamma$.

Thus Γ is nonempty. Let $K \in \Gamma$. As 3 divides the order of K, $Z(K) \in \langle t\rangle^G$, so conjugating in G we may take $\langle t\rangle = Z(K) = Z(K_1)$. Let A and A_1 be Sylow 5-subgroups of K and K_1, respectively. Each is Sylow in H, so conjugating in H we may take $A = A_1 = Z$. Now if (1) holds then $E(C_L(t))$ is conjugate to K_1 in $N_G(Z)$, so $L = \langle Q, E(C_L(t))\rangle$ is conjugate to $L_1 = \langle Q, K_1\rangle$. That is G is transitive on Ω. $|N_G(L):L|$ divides $|N_G(Q):Q| = 32$, so $|N_G(L):L| = 2$. Therefore (1) implies (2), so it remains to establish (1).

From 12.12, $J = O_2(H)$ is of order 2^{11} and $\Phi(J) = E \cong E_{32}$. Let $\bar{H} = H/E$. By 12.12 K acts decomposably on $\bar{J} \cong E_{64}$ with $C_{\bar{J}}(K) = \langle \bar{x}_1, \bar{x}_2\rangle \cong E_4$ and $\bar{J}/C_{\bar{J}}(K)$ the natural module for $K^* \cong L_2(4)$. Let α be an element of order 4 in K and $\bar{I} = \bar{K}\bar{J}$. Then $C_{\bar{J}}(K)\bar{\alpha}$ is fused under $C_{Aut(\bar{I})}(\bar{J})$, so every complement to $\bar{J}$ in $\bar{I}$ containing $\bar{\alpha}$ is in the class $\bar{K}^{\bar{I}}$.

Denote by X the Sylow 3-group of H denoted by Q in Parrott [33]. We may take $X \leq K$ and α to invert X. Let $K_2 \in \Gamma$ with $\langle t\rangle = Z(K_2)$ and $X \leq K_2$, and let $\beta \in$

$\alpha J \cap K_2$. Then β is of order 4 and inverts X, so we may take α and β to lie in the Sylow 2-group T of H described in lemma 12 of [33], since $N_H(X) = XN_T(X)$. The relations for $N_T(X)$ are listed on page 35 of [33]. We find $\langle yx_1 \rangle$ is the unique cyclic subgroup of $N_T(X)$ of order 4 which does not centralize X. Hence we have $\alpha = \beta = yx_1$. In particular $\bar{\alpha} = \bar{\beta}$, so by an earlier remark, $\bar{K}_2 \in \bar{K}^H$. Then as $X \leq K \cap K_2$ and $C_{\bar{J}}(\bar{X}) = 1$, even $\bar{K} = \bar{K}_2$.

Now choose $K \leq L$ and observe K centralizes v of order 4 in sL. Then $v \in C_T(X\langle\alpha\rangle) = \langle x_1 \rangle$, so $v = x_1$. That is $\bar{v} \in C_{\bar{J}}(K) = J_1/E$. By 12.2.2, $J_1 = C_J(E)$. So $\langle v \rangle \trianglelefteq \langle v,E \rangle$. Thus $\langle v \rangle \trianglelefteq \langle v,E,K \rangle$, so $K_2 \leq C_H(v)^{\infty} = KE$. From the relations in lemma 12 of [33], E is an decomposable for K with $Z = C_K(E)$, so there are four KE-classes of complements to $\tilde{E}$ in $\tilde{K}\tilde{E}$. Moreover these relations show these classes fall into two $\langle x_2 \rangle$-orbits. Only one of these orbits is invariant under $u \in N_H((\langle v \rangle)) - I$, so if $\mathrm{Aut}_{N_H(\langle v \rangle)}(K_2) \cong PGL_2(5)$, then K_2 is conjugate to K under $N_H(\langle v \rangle)$. On the other hand $\langle v \rangle = C_J(K_2)$, so as $\mathrm{Aut}_G(K_2) \cong PGL_2(5)$ also $\mathrm{Aut}_{N_H(\langle v \rangle)}(K_2) \cong PGL_2(5)$. Thus the proof of (1) is complete.

Notice that 22.1.2 establishes Theorem 21.4 when G is Ru.

In the remainder of this section let G be Co_1. From the action of G on the Leech lattice, G contains $L \cong U_3(5)$. We may assume $Q\langle t \rangle \leq L$. Then $K = E(C_L(t)) \cong SL_2(5)$. Let $U \in Syl_2(N_K(Q))$.

(22.2) $N_G(Q)/Q \cong GL_2(5)$ and $UQ/Q = Z(N_G(Q)/Q)$.

Proof. A Sylow 2-group R of $N_G(Q) \cong GL_2(5)/Q$ is isomorphic to $Z_4 wr Z_2$, so it has a subgroup R_0 isomorphic to Z_4^2 of index 2 and if $Z_8 \cong V \leq R$ then $C_{R_0}(V) = Z(R) \cong Z_4$. So as $N_L(Q)/Q \cong Z_8$, the lemma holds.

(22.3) t is 2-central in G, so $O_2(H) = P \cong D^4$ and $H^* \cong \Omega_8^+(2)$.

Proof. For some choice of L, $L \leq G_3 \leq G$ with $G_3 \cong Co_3$. Co_3 has two classes of involutions with centralizers $Sp_6(2)/Z_2$ and $Z_2 \times M_{12}$. As $t \in K^{\infty} \leq C_{G_3}(t)^{\infty}$, t is in the first class. So 27 divides $|C_{G_3}(t)|$ and hence also $|C_G(t)|$, which implies the lemma.

(22.4) (1) G is transitive on its $U_3(5)$-subgroups.

(2) $N_G(L) \cong Aut(U_3(5))$.

(3) $\tilde{P}$ is the sum of two $\Omega_4^-(2)$-modules for K^*, so P is transitive on the $SL_2(5)$-subgroups in KP and $N_H(K)^* = N_{H^*}(K^*)$.

(4) The three irreducible K^*-subspaces of the orthogonal space $\tilde{P}$ are totally singular and conjugate under $N_{H^*}(K^*) \cong S_3 \times PGL_2(5)$.

(5) Let $\tilde{I}$ be an irreducible K^*-submodule of $\tilde{P}$ and $H_0 = N_H(\tilde{I})$. Then K^* acts decomposably on $O_2(H_0^*) \cong E_{2^6}$ with $C_{O_2(H_0^*)}(K^*) \cong Z_2$.

(6) H and H_0 are transitive on the G-conjugates of K contained in the respective group. $N_{O_2(H_0)}(Z)$ is transitive on the two conjugates of K contained in $KO_2(H_0)$ and containing Z.

Proof. By 20.3 we can regard $\tilde{P}$ as an 8-dimensional orthogonal space for $H^* \cong \Omega_8^+(2)$. $K^* \cong L_2(4)$ with $C_{\tilde{P}}(Z^*) = 0$ by 22.2. Let $U^* = \langle u^* \rangle$. Then u^* inverts Z^* so $m([\tilde{P},u^*]) = m(\tilde{P})/2 = 4$ as Z^* has no fixed points on $\tilde{P}$. In particular $\tilde{P}$ is transitive on the involutions in $\tilde{U}\tilde{P}$. Hence there are no involutions in H covering u^*.

Next u^* is of type a_4 or c_4 on $\tilde{P}$. As u^* centralizes an $L_2(4)$-subgroup of $N_H(Q)^*$ by 22.2, it must be the former. Let $\tilde{I}$ be an irreducible K-submodule of $\tilde{P}$. If $\tilde{I}$ is nondegenerate then $[\tilde{I},u]$ is not totally singular, contradicting u^* of type a_4. So $\tilde{I}$ is totally singular. Therefore $\tilde{P}/\tilde{I}$ is dual to $\tilde{I}$ as a K^*-module. Also $H_0^* = N_{H^*}(\tilde{I}) \cong \Omega_6^+(2)/E_{64}$ is a maximal parabolic of H^* with $C_H(\tilde{I})^* = O_2(H_0^*)$ the $\Omega_6^+(2)$-module for $\bar{H}_0 = H_0/O_2(H_0)$. Therefore either:

(i) $O_2(H_0^*) = C_{O_2(H_0^*)}(K^*) \oplus [O_2(H_0^*),K^*]$ with $[O_2(H_0^*),K^*]$ the $\Omega_4^-(2)$-module for K^*, and $N_{\bar{H}_0}(\bar{K}) \cong \Gamma L_2(4)$, or

(ii) $O_2(H_0^*)$ is an decomposable for K^* of type 1/4/1 and $N_{\bar{H}_0}(\bar{K}) \cong PGL_2(5)$.

Moreover in each case $\bar{K}$ is determined up to conjugacy in $\bar{H}_0$.

In case (i), $\tilde{I}$ is the $L_2(4)$-submodule for K^* while in case (ii) $\tilde{I}$ is the $\Omega_4^-(2)$-module. This is because $\bar{K}$ centralizes an element of order 3 in case (i).

In case (i), as the $\Omega_4^-(2)$-module is projective, there is one class of complements to $O_2(H_0^*)$ in $K^*O_2(H_0^*)$. Thus K^* is contained in a Levi factor H_1^* of H_0^*. H_1^* stabilizes a complement $\tilde{J}$ to $\tilde{I}$ in $\tilde{P}$, so $\tilde{P}$ is a homogeneous K^*-module. There is an involution $s \in L$ inducing an outer automorphism of K. Then s fixes three of the five irreducible K^*-submodules of $\tilde{P}$, so we may take s to act on $\tilde{I}$ and $\tilde{J}$. Then $[\tilde{P},s] = [\tilde{I},s]\oplus[\tilde{J},s]$ is of rank 4 and as $[\tilde{I},s]$ and $[\tilde{J},s]$ are totally singular, so is $[\tilde{P},s]$. Then s^* is of type a_4 and indeed as $m(\tilde{I}\cap[\tilde{P},s])$ and $m(\tilde{I}\cap[\tilde{P},u])$ are even, u^* and s^* are fused in H^*. This contradicts the first paragraph of this proof, since s is an involution.

Therefore case (ii) holds. Thus $\tilde{I}$, and hence also $\tilde{P}/\tilde{I}$, are $\Omega_4^-(2)$-modules for K^*. So as that module is projective, $\tilde{P}$ is a homogeneous K^*-module with 3 K^*-irreducibles. By an earlier remark each irreducible is totally singular, so as $C_{O_2(H_0^*)}(K^*)$ acts as a transposition on these irreducibles, $C_{H^*}(K^*) \cong S_3$ is transitive on them. Also as the modules are projective, $\tilde{P}$ is transitive on the complements to $\tilde{P}$ in $\tilde{K}\tilde{P}$. Therefore (3), (4), and (5) are established.

H^* has two orbits on maximal totally singular subspaces of $\tilde{P}$, but only one type of subspace is stabilized by z. Thus to show H is transitive on $K^G\cap H$ it suffices to show H_0 is transitive on $K^G\cap H_0$. By (5), if $Z \leq K^g \leq H_0$ then $\bar{K}^g$ is conjugate to $\bar{K}$ under $N_{\bar{H}_0}(\bar{Z})$. Also by (3) and (5) there are two classes of complements to $O_2(H_0)$ in $O_2(H_0)K$. But only one of these classes intersects K^G as members of the other stabilize a nondegenerate 4-subspace of $\tilde{P}$. Therefore (6) is established.

By (6), K is determined up to conjugacy in $N_H(Z)$, so $L = \langle Q,K\rangle$ is determined up to conjugacy in G. Hence (1) holds. $N_H(K)$ acts on $\langle Q,K\rangle = L$, so as $N_H(K)^* = N_{H^*}(K^*) \cong S_3 \times S_5$, (2) holds.

(22.5) (1) L is contained in subgroups G_i of G, $1 \leq i \leq 6$, with $G = G_1$, $G_2 \cong Co_2$, $G_3 \cong Co_3$, $G_4 \cong Mc$, $G_5 \cong HS$, and $G_6 = N_G(L)$.

(2) G is transitive on its subgroups isomorphic to G_i for each i, $1 \leq i \leq 5$.

(3) G_i is transitive on its $U_3(5)$-subgroups for i = 1,2, and 4.

(4) G_5 has two classes of $U_3(5)$-subgroups fused in $Aut(G_5)$.

(5) G_3 has two classes of $U_3(5)$-subgroups with representatives L_1 and L_2. $N_{G_3}(L_1) \cong Aut(U_3(5))$ and is maximal in G_3. $N_{G_3}(L_2) \cong Z_2/U_3(5)$ and is contained in subgroups of G_3 isomorphic to HS and Aut(Mc).

(6) G_2 and G_3 are transitive on their subgroups isomorphic to Mc and HS.

(7) $N_{G_i}(L) \cong Z_2/U_3(5)$ for i = 2 and 5 while $N_{G_4}(L) = L$.

(8) The normalizers of subgroups of G_2 and G_3 isomorphic to HS and Mc are described in Table 21.4.

Proof. The fact that G contains subgroups of the prescribed isomorphism type and that G_i contains a $U_3(5)$-subgroup can be seen from the action of G on the Leech lattice. So by 22.4.1 we may take $L \leq G_i$. Hence (1) holds. It will take longer to prove the remaining parts.

Let $U \leq V \in Syl_2(N_L(Q))$. Then $V \cong Z_8$. Let $X = N_G(Q)$, $X\alpha = X/Q$, and $X_i = G_i \cap X$. From section 5 in [33]:

(22.6) (1) $X\alpha \cong GL_2(5)$.

(2) $(X_2)\alpha \cong Z_2/(Z_4 * SL_2(3))$.

(3) $(X_3)\alpha \cong Z_2/Z_{24}$.

(4) $(X_4)\alpha \cong Z_2/Z_{12}$.

(5) $(X_5)\alpha \cong Z_2/Z_8$.

(6) $(X_6)\alpha \cong Z_2/Z_{24}$.

We conclude from 22.6 that $(X_i)\alpha = N_{X\alpha}(B_i)$ for some $B_i \in Syl_i(E(X\alpha))$ if i = 2 or 3, $(X_5)\alpha \in Syl_2(N_{X\alpha}(V\alpha))$, and $(X_6)\alpha = N_{X\alpha}(V\alpha)$. Also as $V \cong X_4$, $(X_4)\alpha = V\alpha B$ for some $B \in Syl_3(E(X\alpha))$ with $B = [V\alpha, B]$. In particular:

(22.7) (1) If i is not 3 or 6 then $(X_i)\alpha$ is determined up to conjugacy under $N_{X\alpha}(V\alpha)$ by its isomorphism type and the constraint $V\alpha \leq (X_i)\alpha$.

(2) $(X_6)\alpha = N_{X\alpha}(V\alpha)$.

(3) $N_{X\alpha}(V\alpha)$ has two orbits on subgroups of $X\alpha$ isomorphic to $(X_3)\alpha$ containing $V\alpha$. $V\alpha$ is normal in members of one orbit but not the other.

Next we conclude from 22.6 and Theorem B that:

(22.8) One of the following holds:

(1) $G_i = \langle X_i, L \rangle$.

(2) i = 3, $VQ \trianglelefteq X_i$, and $N_G(L) = \langle X_i, L \rangle$.

(3) i = 5 and $X_i \leq N_G(L)$ with $|\langle X_i, L \rangle : L| = 2$.

As $N_G(QV) \leq N_G(L)$ and L is transitive on its subgroups isomorphic to QV, we conclude from 22.4.1 that:

(22.9) L is the unique $U_3(5)$-subgroup of G containing QV.

(22.10) There is an outer automorphism β of G_5 acting on Q with $G_5 = \langle L,L^\beta\rangle$.

Proof. We can choose β to be an outer automorphism of G_5 acting on Q with $(QV)^\beta \neq QV$ as V is not normal in a Sylow 2-group of $N_{G_5}(Q)\langle\beta\rangle$. So $L \neq L^\beta$ by 22.9. Then $G_5 = \langle L,L^\beta\rangle$ by Theorem B.

(22.11) G_3 has at least two classes of $U_3(5)$-subgroups with representatives L_1 and L_2 such that $N_G(L_1) = N_{G_3}(L_1)$ and $|N_{G_3}(L_2):L_2| = 2$.

Proof. We observed during the proof of 22.3 that $H \cap G_3 = H_3 \cong Sp_6(2)/Z_2$. Thus either H_3^* is the stabilizer in H^* of a nonsingular point of $\tilde{P}$ or H_3^* acts irreducibly on $\tilde{P}$ in the spin representation. As $K^* \leq H_3^*$ fixes no point of $\tilde{P}$ it must be the latter, and indeed $H_{30} = H_0 \cap G_3 \cong S_6/E_{32}/Z_2$. By 22.4.5 there are just two conjugates K_1^* and K_2^* of K^* under $O_2(H_0^*)$ containing Z^*, and as H_{30}^* contains both of these conjugates we may take $K_i \leq H_{30}$ by 22.4.3. As $C_{O_2(H_{30}^*)}(Z^*) \leq N(K^*)$, K_1 is not fused to K_2 in H_{30}. A Levi factor J^* of H_{30}^* stabilzes a complementary conjugate $\tilde{I}^g$ of $\tilde{I}$ under H_3 and hence is contained in a Levi factor of H_0^*. From this we see that J contains K_1 or K_2, say K_1. Now as K_1 centralizes a subgroup of order 2 in $O_2(H_{30}^*)$ and $O_2((H_{30}^*)^g)$ it follows that $N_{H_3^*}(K_1^*) \cong S_3 \times PGL_2(5)$ and is transitive on the H_3-conjugates of H_0 containing K_1. The first remark says that $N_{G_3}(L_1) = N_G(L_1)$, where $L_1 = \langle K_1,Q\rangle$. The second says H_{30} is transitive on the H_3-conjugates of K_1 contained in H_{30}, so that K_2 is not H_3-conjugate to K_1. Thus from 22.7, $|N_{G_3}(L_2):L_2| = 2$, where $L_2 = \langle K_2,Q\rangle$.

(22.12) If $QV \leq M \leq G$ with $M \cong G_i$ for some i, then $L \leq M$.

Proof. We have seen there is $L_0 \leq M$ with $Q \leq L_0 \cong U_3(5)$, and that $N_{L_0}(Q) = QV_0 \in (QV)^X$. By 22.6 either QV_0 is fused to QV in X or i = 3 or 5. In the first case $L \leq M$ by 22.9, so assume the second. Then by 22.10 and 22.11 there are at least two

M-classes of $U_3(5)$-subgroups of M, while by 22.6 there are just two $X \cap M$-orbits on $(QV)^X \cap M$. Thus even in this case we may choose L_0 so that QV_0 is fused to QV in $X \cap M$, completing the proof.

It follows from 22.9 and 22.12 that the map $QV^x \mapsto L^x$ is a bijection of $(QV)^X \cap G_i$ with $L^X \cap G_i$ commuting with the permutation actions of X_i on the respective sets. This observation together with 22.10 and the action of X_i on $(QV)^X \cap X_i$ imply parts (3), (4), and (7) of 22.5, and the corresponding portion of part (5). These remarks together with 22.8 show

$$G_i = \langle L^x : x \in X, V^x \leq X_i \rangle$$

for each i, $1 \leq i \leq 5$. Then 22.7 implies 22.5.2. Similar remarks give parts (6) and (8) and complete the proof of part (5) of 22.5.

So 22.5 is at last established. It is not difficult to see that 22.5 completes the proof of Theorem 21.4.

23. Nonlocal overgroups of large symplectic subgroups.

In this section G is a finite group, Q is a large symplectic p-subgroup of G of width at least 2, and $\langle z\rangle = Z = \Omega_1(Z(Q))$. Assume $F^*(G)$ is simple and exclude case (2) of Theorem A. We prove:

Theorem 23.1. Table 23.1 lists those M in $\mathcal{N}(Q)$ with $F^*(M) \neq O_p(M)$. Column 1 lists Q, column 2 lists G, column 3 lists the isomorphism type of M, column 4 lists the number of conjugates of M containing a Sylow p-subgroup of M, and column 5 lists the number of orbits of G on subgroups isomorphic to M. All orbits are fused in Aut(G).

Actually we will completely determine the subgroups of G containing Q whose generalized Fitting subgroup is not a p-group.

Assume for the rest of this section that $L = \langle Q^L\rangle \leq G$ with $F^*(L)$ simple. Then G and L are described in Theorem B. We consider each of the cases arising there. Let $H = N_G(Z)$, $H^* = H/Q$, and $\tilde{H} = H/Z$. For $X \leq G$ let $X_L = X\cap L$. Let $T \in Syl_p(H)$ with $T_L \in Syl_p(L)$. Observe that the entry in column 4 is just $|N_H(T_L):N_{H_L}(T_L)|$, which is easily calculated.

Assume first that p = 2. The cases $G \cong P\Omega_6^{\epsilon}(3)$, $G_2(3)$, and $P\Omega_8^{+}(3)$ are handled in 9.8. The case $G \cong Co_1$ and $L \cong Co_2$ is handled in 22.5.2. The case $G \cong Co_1$ or Co_2 and $L \cong U_6(2)$ follows as a corollary to the corresponding case for p = 3, which will be dealt with later. The case $G \cong M(24)'$ and $L \cong \Omega_{10}^{-}(2)$ is handled in 15.9.5. Finally the case $G \cong F_3$ and $L \cong {}^3D_4(2)$ is handled in 15.16.3.

Assume Q is 3^{1+8} and G is F_2. Then $H^* \cong O_6^{-}(2)/Q_8^3$. Further $L \cong M(23)$ or $P\Omega_8^{+}(3)$ with $O^{3'}(H_L^*) \cong GU_3(2)/Q^3$ or $\Omega_4^{+}(3)*SL_2(3)$, respectively. Let $R = O_{3,2}(H)$ and $P^* = J(T^*) \cong E_{27}$. Then $R \leq H_L$ and as $T_L \in Syl_3(L)$, $P \leq H_L$. Indeed if $L \cong P\Omega_8^{+}(3)$ then $O^{3'}(H_L) = RP$ while if $L \cong M(23)$ then $O^{3'}(H_L)/R = O^{3'}(X/R)$, where X/R is the maximal parabolic of H/R (regarded as the group $GPSp_4(3)$ of characteristic 3) containing T/R and not normalizing PR/R. So in either case $O^{3'}(H_L)$ is uniquely determined by T_L.

Table 23.1

Q	G	M	Conjugates over S	Orbits on M-subgroups
D^2	$U_4(3)$	$U_4(2)$	2	2
D^2	$L_4(3)$	$O_6^-(2)$	1	2
D^2	$G_2(3)$	$G_2(2)$	1	2
D^4	$P\Omega_8^+(3)$	$\Omega_8^+(2)$	1	4
D^4	Co_1	Co_2	4	1
		$Aut(U_6(2))$	8	1
D^4	Co_2	$Z_2/U_6(2)$	4	1
D^4	F_3	$Aut(^3D_4(2))$	8	1
D^6	$M(24)'$	$\Omega_{10}^-(2)$	2	1
3^{1+8}	F_2	$Aut(P\Omega_8^+(3))$	1	1
		$M(23)$	1	1
3^{1+8}	$M(23)$	$S_3/P\Omega_8^+(3)$	1	1
3^{1+6}	$^2E_6(2)$	$M(22)$	1	3
3^{1+6}	$M(22)$	$P\Omega_7(3)$	1	2
3^{1+4}	Co_1	Co_2	9	1
		Co_3	3	1
		$Aut(U_6(2))$	3	1
3^{1+4}	Co_2	Mc	4	1
		$Aut(U_4(3))$	1	1
		$Z_2/U_6(2)$	2	1
3^{1+4}	Co_3	$Aut(Mc)$	6	1
		$E_4/U_4(3)$	3	1
3^{1+4}	Mc	$U_4(3)$	2	1
3^{1+4}	$U_6(2)$	$Z_2/U_4(3)$	1	3
3^{1+4}	$F_4(2)$	$Z_2/L_4(3)$	1	1
		$Aut(\Omega_8^+(2))$	2	2
5^{1+4}	F_2	$Aut(F_5)$	1	1
5^{1+4}	Ly	$G_2(5)$	1	1

Let $g \in L$ with $z^g \in \mathcal{Y}(T_L) \subseteq \mathcal{Y}(T)$; this choice is possible by 19.8. $O^{3'}(H_L)$ is irreducible on $\tilde{Q}$ so by 20.1 and Theorem B, $L = \langle O^{3'}(H_L), Q^g\rangle$. Thus there is at most one subgroup of G isomorphic to L with $T_L \in Syl_3(L)$. Moreover $N_G(L) = LN_H(T_L)$ as T_L is contained in a unique conjugate of H.

Let $M = N_G(L)$. Then $L = F^*(M)$ so if $L \cong M(23)$ then $M = L$ since $L = Aut(L)$. On the other hand if L is $P\Omega_8^+(3)$ then $|N_H(T_L)O^{3'}(H_L)| = |C_A(Z)|$, where $A = Aut(L)$, so $N_G(L) \cong A$.

By 16.8.5 $N_G(L_1) \not\leq L_2$ for $L_1 \cong P\Omega_8^+(3)$ and $L_2 \cong M(23)$. So each is maximal in G by Theorem B. Finally from section 15, F_2 contains a copy of M(23) while by 16.8 M(23) contains a copy of $P\Omega_8^+(3)$, so the proof of Theorem 23.1 is complete when G is F_2.

Most of the remaining cases are similar. When this is so the details which go smoothly are omitted. For example if G is M(23) or G is F_2 and Q is 5^{1+4}, the proof outlined above goes through with no problem and all details are omitted. (Or appeal to 16.8 when G is M(23).)

Let G be ${}^2E_6(2)$ so that Q is 3^{1+6} and $R = O_{3,2}(H)$ is the central product of 3 copies R_i, $1 \leq i \leq 3$, of $SU_3(2)$, permuted transitively by H. H/R is E_9 inverted by an involution. By Theorem B, L is M(22) or $P\Omega_7(3)$, with $|H_L^*| = 2^9.9$ or $H_L^* \cong Z_2/(\Omega_3(3) \times SL_2(3))$, respectively. Let $\bar{R} = R^*/Z(R^*)$. Then $\bar{R} = \bigoplus_{i=1}^{3} C_{\bar{R}}(D_i^*)$, where D_i^*, $1 \leq i \leq 3$, are the three subgroups of T^* of order 3 moving the subgroups R_i. Moreover $C_{R^*}(D_i^*) \cong Q_8$. Hence there are exactly three subgroups of H^* of order $2^9.9$ containing $T = T_L$, and these are $N_H(T)^*C_R{}^*(D_i)C_R{}^*(D_j)$, as {i,j} varies over the subsets of {1,2,3} of order 2. $N_{Aut(G)}(Q)$ permutes these 3 subgroups as S_3 and H_L^* is one of them if L is M(22). Similarly there are three subgroups of H^* isomorphic to H_L^* when L is $P\Omega_7(3)$. These subgroups are $N_H(T)^*C_{R^*}(D_i)Z(R^*)$, $1 \leq i \leq 3$, which are permuted as S_3 by $N_{Aut(G)}(Q)$. With these remarks the proof of Theorem 23.1 goes through as above for $G \cong {}^2E_6(2)$ or M(22).

Next let $G \cong Ly$. Then $Q \cong 5^{1+4}$ and by Theorem B, L is $G_2(5)$. $H^* \cong Z_2/(Z_4*SL_2(9))$, so T is contained in exactly two subgroups K_1 and K_2 with $K_i^* \cong Z_2/(Z_4*SL_2(5))$. Let $z^g \in \mathcal{Y}(T)$. On pages 561-562 of [42], Lyons uses the Brauer trick to show that, for suitable choice of notation, $\langle K_1, Q^g\rangle = M_1$ is a proper subgroup of G and $z \notin O_5(M_1)$. Hence by 17.4 and Theorem B, $M_1 \cong G_2(5)$. Then to complete the proof of Theorem 23.1 when G is Ly it suffices to show:

(23.2) Let $Z^g \in \mathcal{Y}(T)$ and $V = Q\cap Q^g$. Then $N_G(V) = \langle K_2, Q^g\rangle$ with $N_G(V)/V \cong L_3(5)$.

Let $X = \langle Q, Q^g\rangle$. From 17.8, 19.2, and 19.7, $V \cong E_{5^3}$ and $X/C_X(V) \cong SL_2(3)$ acts naturally on $U = \langle z, z^g\rangle$. Thus if t is an involution in X, $U = [V,t]$ and $W = C_V(t)$ is of order 5. Then $C_X(W)$ involves $SL_2(5)$, so as the centralizer in G of a non-5-central element is solvable, $W = Z^h \in Z^G$. Also $C_X(W)$ is irreducible on V/W and by 17.5, $z \in Q^h$, so $V \leq Q^h$. Thus $V = Q\cap Q^h$. Now 23.2 follows easily from 17.6.

We see from Theorem B that we have reduced to the case $Q \cong 3^{1+4}$. The situation here is a little different from the proceeding cases. For one thing by 19.8, either $\mathcal{Y}(T)$ is empty or G is $F_4(2)$, and in the latter case T is not transitive on $\mathcal{Y}(T)$.

Let G be $F_4(2)$ and $A = \mathrm{Aut}(G)$. Then $|A:G| = 2$ and we let $H_A = N_A(Z)$ and define H_A^*, $\tilde{H}_A$, as above. Let $V = J(T)$. By 19.6, $V \cong E_{81}$ and $O^{3'}(N_G(V)/V) \cong \Omega_4^+(3)$ acts naturally on V with $Z^G\cap V$ the set of singular points in the orthogonal space V. Hence $(Z_1)^T \cup (Z_3)^T = \mathcal{Y}(T)$ is of order 6 with $Z_i = Z^{g_i} \leq V$. Let $V_i = Q\cap Q^{g_i}$. We construct a subgroup isomorphic to $L_4(3)$, and identify that subgroup via its Tits geometry.

Let $R = O_{3,2}(H)$. Then $R = R_1 * R_2$ is the central product of copies R_i of $SU_3(2)$ permuted transitively by H_A, with $H/R \cong S_3$ and $H_A/R \cong Z_2 \times S_3$. In particular H_A has two orbits on the set of subgroups of Q of order 3 distict from Z, those orbits are of order 3.32 and 3.8, and the first orbit consists of those subgroups not in R_1 or R_2. If $Z_i \leq R_j$ then aas $H/R \cong S_3$, $\Phi(R_{3-j}) \leq R^{g_i}$ and hence $O_3(R_{3-j}) \leq Q^{g_i}$. So $Z = \Phi(R_{3-j}) \leq \Phi(Q^{g_i}) = Z_i$, a contradiction. So $Z^G\cap Q$ is of order (3.32)+1 and Z is weakly closed in R_j with respect to G.

We next observe that $O^{3'}(N_G(V_j)/V_j) \cong L_3(3)$ acts naturally on V_j. First as $|Q^{g_j}\cap H|$ 3^4 and $|H/Q| = 3$, $V_j \cong E_{27}$. Then using the same argument which established 23.2, it suffices to show $(V_i)^{\#} \subseteq Z^G$. Let $\langle t\rangle = Z_1$. Then $V_1 \leq C_Q(t) = \langle t_1, t_2, s_1 s_2, z\rangle$ where $\tilde{t}_i$ is the projection of $\tilde{t}$ on $\tilde{Q}_i = O_3(\tilde{R}_i)$ and $Q_i = \langle t_i, s_i\rangle$. In particular the members of $C_Q(t)^{\#}$ not in Z^G are in $\langle t_i, z\rangle - Z$. However $V\cap Q = \langle t_1, t_2, z\rangle$ and $V_1\cap V = \langle t, z\rangle$, so the observation is established.

Let $M_i = O^{3'}(N_G(V_i))$, $i = 1,3$. Let s be an involution in R_1 and observe $(\tilde{Z}_1)^s = \tilde{Z}_3$, so $(V_1)^s = V_3$ and hence $(M_1)^s = M_3$. Let $M_{13} = M_1 \cap H$. Then $M_{13}^* \cong GL_2(3)$ and $[s^*,H^*] \leq Z(M_{13}^*)$, so $s \in N_G(M_{13})$. Thus $M_{13} \leq M_1 \cap (M_1)^s = M_1 \cap M_3$, and as M_{13} is maximal in M_1, even $M_{13} = M_1 \cap M_3$.

Next $N_G(V)/V$ is isomorphic to the Weyl group of F_4 and hence is $O_4^+(3)$ acting naturally on V. In particular $N_G(V)$ has two subnormal subgroups Y_1 and Y_3 with $Y_i/V \cong SL_2(3)$. Let M_2 be the subgroup of $N_G(V)$ of index 2 fixing Y_1 and Y_3 and let $M_{i2} = N_{M_i}(V)$ for $i = 1,3$. $M_{j2}/V \cong Z_2/(SL_2(3) \times Z_3)$, so $O_{3,2}(M_{j2}) = O_{2,3}(Y_{i(j)})$, for some $i(j)$ depending on j. As $\langle M_{13}, O_{2,3}(M_{12})\rangle = M_1 \neq M_3 = \langle M_{13}, O_{2,3}(M_{32})\rangle$, we may take $i(j) = j$ for $j = 1,3$. Then $M_{j2} \leq N_G(V) \cap N(Y_j) \leq M_2$ and indeed $M_2 = M_{12}M_{32}$.

Let $M = \langle M_i : 1 \leq i \leq 3\rangle$. By the discussion above and section 27, M preserves a Tits geometry over GF(3) with Coxeter diagram of type A_3, so $M \cong L_4(3)$ (cf Proposition 6 in [38].) Moreover $(M_1)^s = M_3$ and $M = \langle M_1, M_3\rangle$, so s induces an outer automorphism on M nontrivial on the Dynkin diagram of M.

We have shown G possesses a subgroup $M \cong Z_2/L_4(3)$. On the other hand if $L \cong L_4(3)$ then $L = \langle N_L(V_1), N_L(V_3)\rangle \leq M$, so $L = F^*(M)$ is unique. By Theorem B, $M \in \mathcal{N}$ and indeed either $L = F^*(M)$ or $L \cong Z_3/\Omega_8^+(2)$. From 4.1 in [23] there exists $Z_3/\Omega_8^+(2) \leq G$. It remains to show there are exactly two choices for L, subject to $T \leq L \cong Z_3/\Omega_8^+(2)$, and these two subgroups are fused in $N_A(T)$.

Observe first that $R_i \trianglelefteq H$ for $i = 1,2$. For if u is an involution in $Z^*(N_G(V))$ then as u inverts $Q \cap V$, u inverts z, so $u \in H - O^2(H)$ and u inverts $1 \neq \tilde{V} \cap \tilde{Q}_i$. Hence u acts on R_i. Therefore the observation is established.

Next $O_{3,2}(H_L) \cong SU_3(2)*3^{1+2}$, so $O_{3,2}(H_L) = R_iQ$ for $i = 1$ or 2. Hence as $L = \langle O_{3,2}(H_L), O^{3'}(N_G(V))\rangle$ and H_A permutes R_1 and R_2 transitively, the proof of Theorem 23.1 is complete in the case $G \cong F_4(2)$.

Finally let $G = G_1 \cong Co_1$, $G_2 \cong Co_2$, $G_3 \cong Co_3$, $G_4 \cong Mc$, and $G_6 \cong U_6(2)$. We saw in the last section that we may take $G_i \leq G$ for $i \leq 4$. From the action on G on the Leech lattice, G contains a copy of $U_6(2)$ which we may take to be G_6. Of course $U_6(2)$ contains a copy of $U_4(3)$ so there is $U_4(3) \cong L \leq G$.

Let $X = O^{3'}(H_L)$. Then $X^* \cong SL_2(3)$ and $B = N_{Aut(L)}(Q)/Q \cong Z_2/(D_8*SL_2(3))$. Now $X^* \leq B \leq Out(Q) \cong GSp_4(3) \cong H^*$. Further if $E = Z(H^*)$ then B/E is contained in a 2-parabolic D of $H^*/E \cong O_6^-(2)$ and as an element of order 3 in B/E centralizes a subgroup of order 24 in B/E it follows that $D \cong S_5/E_{16}$ and $O^2(B)/E$ is determined up to conjugacy in D and hence also in H^*/E. Thus $X^* = O^2(B)$ is determined up to conjugacy in H^*.

Let $P \in Syl_3(X)$. Then $J(P) = J \cong E_{81}$ and $I = J \cap Q \cong E_{27}$. So $\tilde{I}$ is a maximal totally singular subspace of $\tilde{Q}$ and hence determined up to conjugacy under H. In particular from 19.3 there exists $T \in Syl_3(H)$ with $I = V \cap Q$, where $V = C_G(I) = J(T) \cong E_{3^6}$ and $N_G(V)/V \cong M_{12}/Z_2$. Then $J \leq C_G(I) = V$ and $N_G(J) = C_G(J)(N_G(J) \cap N_G(V))$ by a Frattini argument. But $C_G(J) \leq C_G(I) = V$, so $N_G(J) \leq N_G(V)$.

$N_L(J)/J \cong \Omega_4^-(3)$ acts naturally on J and as $|V:J| = 9$, $N_L(J) = N_L(J)^\infty$ centralizes V/J. Thus $N_L(J) = (N_L(J)V)^\infty$. As $N_G(V)/V \cong M_{12}/Z_2$, we conclude from the table of maximal subgroups of M_{12} in [19] that if $N_L(J)V \leq Y \leq N_G(V)$ then either $N_L(J) \trianglelefteq Y$ or $(Y/V)^\infty \cong M_{11}$ or M_{12}/Z_2. Neither group is contained in $GL_4(3)$ so $N_L(J)V \trianglelefteq N_G(J)$. Therefore $N_L(J) = N_G(J)^\infty$ is determined by J and hence $L = \langle X, N_L(J)\rangle$ is determined by X. Thus as X is determined up to conjugacy in H, so is L subject to $Q \leq L$.

Also $N_G(L) = LN_H(L)$ and $N_H(L) = N_H(X)$. As $N_H(X)^* \cong N_{Aut(L)}(X)/Q$ we conclude $N_G(L) \cong Aut(L)$.

We have shown:

(23.3) $G \cong Co_1$ has a unique conjugacy class of subgroups L isomorphic to $U_4(3)$. $N_G(L) \cong Aut(L)$.

(23.4) $L \leq G_i$ if and only if $X = O^2(H \cap L) \leq H_i = H \cap G_i$.

For if $i \neq 1$ or 3 and $X \leq G_i$ then by 19.6, $JQ \in Syl_3(H_i)$ and $O^2(N_{G_i}(J)) = N_L(J)$, so $L = \langle X, N_L(J)\rangle \leq G_i$. If $i = 1$ the result is trivial, while if $i = 3$ then $J(T \cap G_3) = V \cap G_3$ and an earlier remark shows $N_L(J) \leq N_{G_3}(V) \leq N_{G_3}(J(T \cap G_3))$ so again $L = \langle X, N_L(J)\rangle \leq G_3$.

(23.5) The map $\alpha : D \mapsto H \cap D$ is an injection from the set Δ of subgroups D of G with $Q \leq D$ and $F^*(D) \neq O_3(D)$ into the set Γ of subgroups of H containing an H-conjugate of X. α^{-1} maps Y into $\langle Y, N_G(J^g)^\infty\rangle$, where $X^g \leq Y$ and $J = J(S)$ for $S \in Syl_3(X)$. Moreover for $D,E \in \Delta$:

(1) $D \leq E$ if and only if $D\alpha \leq E\alpha$, and

(2) $D \in E^G$ if and only if $D\alpha \in (E\alpha)^H$.

Observe first that Δ is the set of subgroups satisfying the defining hypothesis of G_i or L and hence $H \cap D$ corresponds to H_i or H_L. Also H_i does contain some conjugate of X, which we may take to be X, and therefore by 23.4, $L \leq G_i$. This shows α does map Δ into Γ. Next by 20.2, H_i is maximal in G_i, so $G_i = \langle H_i, L \rangle$. This shows α is injective, and since we've seen $L = \langle X, N_G(J)^{\infty} \rangle$, the right inverse of α is as claimed. The remaining remarks are easy.

It is now easy to complete the proof of Theorem 23.1 using the correspondence of 23.5. Some parts of the Theorem are already known from 21.4. The proof of the following remark is also contained in the discussion above:

(23.6) Let $G \cong Co_1$ or Co_3, $Q \cong 3^{1+4}$, and A an elementary abelian 3-subgroup of G with $m(A \cap Q) = 3$ and with $A \not\leq Q$ if $G \cong Co_3$. Then there exists $T \in Syl_3(H)$ with $A \leq J(T)$ and $N_G(A) \leq N_G(J(T))$.

It is also worth noting that $G_2 = Co_2$ has two classes $(L_i)^{G_2}$ of $U_4(3)$-subgroups. $N_{G_2}(L_1)$ is maximal and $|N_{G_2}(L_2):L_2| = 2$. L_2 is maximal in G_4, $N_{G_2}(L_2) \leq G_6$, and $|N_{G_2}(L_1):N_{G_2}(L_1) \cap N_{G_2}(G_6)| = 2$.

24. Small Sylow 3-groups.

In this section G is a sporadic group, $Q \in Syl_3(G)$, $Q \cong 3^{1+2}$, $\langle z \rangle = Z = Z(Q)$, and L is a component of $N_G(Z) = H$. Set $\tilde{H} = H/Z$. From section 5 in [23] we find:

(24.1) $L = F^*(H)$ and one of the following holds:

(1) $G \cong M_{24}$ and $\tilde{H} \cong S_6$.

(2) $G \cong J_2$ and $\tilde{H} \cong PGL_2(9)$.

(3) $G \cong J_4$ and $\tilde{H} \cong Aut(M_{22})/Z_2$.

(4) $G \cong He$ and $\tilde{H} \cong S_7$.

(5) $G \cong Ru$ and $\tilde{H} \cong Aut(A_6)$.

(24.2) Either

(1) $C_G(Q) = Z$, or

(2) $G \cong J_4$ and $C_G(Q) = Z(L) \cong Z_6$.

Proof. This follows from 23.1.

(24.3) Let $K = \langle Q^K \rangle \leq H$. Then one of the following holds:

(1) $K = Q$.

(2) $K = L$.

(3) $G \cong J_4$ and $\tilde{K}$ is isomorphic to A_6/E_{32}, A_7, $L_3(4)/Z_2$, A_6, E_9/E_{16}, or $Z_3 \times A_4$.

(4) $G \cong He$ and $\tilde{K}$ is isomorphic to A_6 or $Z_3 \times A_4$.

Proof. This follows from 24.1 and the subgroup structure of $\tilde{L}$.

(24.4) Let Δ be the set of subgroups D of Q of order 9 with $D^\# \subseteq z^G$. Then $N_G(Q)$ is transitive on Δ and $|\Delta|$ is listed in Table 24.4. G has at most two classes of subgroups of order 3, and if X^G is a class distinct from Z^G then $N_G(X)$ is listed in Table 24.4.

Table 24.4

G	M_{24}	J_2	J_4	He	Ru
$\lvert\Delta\rvert$	2	0	4	2	4
$N_G(X)$	$S_3 \times L_3(2)$	$S_3 \times A_4$	—	$S_3 \times L_3(2)$	—

(24.5) Let $S \in \mathcal{U}_G(Q,3')$. Then

(1) S is a 2-group.

(2) If $G \cong M_{24}$ or J_2 then $S = 1$.

(3) If $G \cong Ru$ then $C_S(Z) = 1$.

(4) If $G \cong He$ then either $C_S(Z) = 1$ or $E_4 \cong S \leq H$.

(5) If $G \cong Ru$ or He and $S \neq 1 = C_S(Z)$ then $S \cong E_{64}$.

Proof. By 24.1 and 24.3:

(24.5.6) $C_S(Z)$ is a 2-group.

(24.5.7) Either $C_S(Z) = 1$ or $G \cong J_4$ or He.

Let D be a subgroup of Q of order 9, $\Gamma_1(D)$ the set of subgroups of D of order 3, and $\Gamma(D) = \Gamma_1(D) - \{Z\}$.By 24.4 either we may choose $D \in \Delta$ or $G \cong J_2$ and $N_G(X) \cong S_3 \times A_4$ for each $X \in \Gamma(D)$. Of course $C_S(X) \in \mathcal{U}_{C(X)}(D,3')$ for each $X \in \Gamma_1(D)$. Also $\mathcal{U}_{C(X)}(B,3')$ is a 2-group for each $E_9 \cong B \leq C_G(X)$ by 24.1 and the choice of D, unless G is He or J_4, where $\mathcal{U}_{C(X)}(B,7)$ may be nonempty. We show S cannnot be a nontrivial 7-group; hence $C_S(X)$ is a 2-group for all $X \in \Gamma_1(D)$, so as $R = \langle C_R(X) : X \in \Gamma_1(D)\rangle$, for each D-invariant Sylow r-group R of S and each prime r, (1) will be established.

If S is a 7-group then by 24.5.6, $C_S(Z) = 1$. We may take S abelian, so $S = \bigoplus_{X \in \Gamma(D)} C_S(X)$ is of rank $3m(C_S(X)) \geq 3$. As $m_7(G) \leq 2$ this is impossible.

Suppose $G \cong He$ and $U = C_S(Z) \neq 1$. Then as $U \in \mathcal{U}_H(Q,2)$, $U \cong E_4$ is determined up to conjugacy. If u is a non-2-central involution in G then $U_0 = O_2(C_G(u)) \cong E_4$ with $N_G(U_0) \cong S_3/L_3(4)/E_4$. Thus as U is determined up to conjugacy under $N_G(Q)$, we may take $U = U_0$. But then $U \in \mathcal{U}^*_{N(U)}(Q,2)$, so $S = U$. Thus (4) is established.

To prove the rest of the lemma we may assume G is not J_4 and if G is He, we may take $C_S(Z) = 1$. Hence in each case to be considered, $C_S(Z) = 1$ by 24.5.7. In particular (3) is established.

Let $R = \Omega_1(Z(S))$. Then R is abelian and $C_R(Z) = 1$, so $R = \bigoplus_{X \in \Gamma(D)} C_R(X)$. In particular as Q is transitive on $\Gamma(D)$, $m(R) = 3m(C_R(X))$. $C_S(X) = [C_S(X),Z]$ and by choice of D, if $E \in \mathcal{U}_{C(X)}(D,2)$ with $E = [E,Z] \neq 1$, then $E \cong E_4$. It follows that $R = S \cong E_{64}$, so (5) holds. Finally $m_2(G) < 6$ if G is M_{24} or J_2, so (2) is established.

(24.6) Let $Q \leq M < G$ and $M^* = M/O_2(M)$. Then one of the following holds:

(1) $Z^* \trianglelefteq M^*$.

(2) $O^{3'}(M^*) \cong SL_2(9)/E_9$ and $F^*(M^*) = O_3(M^*)$.

(3) G is M_{24} or J_4 and $F^*(M^*) \cong M_{12}$.

(4) G is J_4 and $M^* \cong M_{24}$.

(5) G is Ru and $F^*(M^*) \cong L_3(3)$ or ${}^2F_4(2)'$.

(6) $F^*(M^*) \cong U_3(3)$ and G is not M_{24} or He.

(7) $O^{3'}(M^*) \cong PGL_3(4)$ and $F^*(M^*) \cong L_3(4)$.

Proof. If $F^*(M^*) = O_3(M^*)$ then (1) or (2) holds, so assume otherwise. By 24.5, $O_{3'}(M^*) = 1$. Then as $3^{1+2} \cong Q^* \in Syl_3(M^*)$, $F^*(M^*) = I^*$ is a nonabelian simple group with $z \in I$. Without loss $M = QI$.

If I^* is sporadic or alternating, then $Q^* \in Syl_3(I^*)$, so I^* is sporadic and either $I^* \cong M_{12}$ or I^* is listed in 24.1. As $|I^*|$ divides $|G|$, either (3) or (4) holds or G is He or Ru and $I^* \cong J_2$. Assume one of the last two cases holds, and let $B \in Syl_5(I)$. From section 5 in [23], $Aut_I(B)$ is irreducible on B and there exists $Z_5 \cong Y \leq B$ with $N_I(B)$ not solvable. But if G is Ru then $B \leq R \in Syl_5(G)$ with $N_G(Z(R))$ solvable. Thus $Y \notin Z(R)^G$ and all subgroups of order 5 in B distinct from Z(R) are fused to Y in R. This is impossible as $N_G(B)$ is irreducible on B by an earlier remark.

So G is He. Claim $O_2(M) = 1$. Assume not and let $A = \Omega_1(Z(O_2(M)))$. $C_G(A)$ does not involve J_2 for any involution a in G, so $O_2(M) = C_M(A)$. Thus $m(A) > 6 = m(G)$, a contradiction.

Therefore $I \cong J_2$. Let t be a 2-central involution in I. Then $C_I(t) \cong A_5/QD$. As 5 divides $|C_G(t)|$, t is not 2-central in G, so $Y = C_G(t) \cong Z_2/L_3(4)/E_4$. Thus $C_I(t)O_2(Y)/O_2(Y)$ is contained in a maximal parabolic $P/O_2(Y)$ of $Y/O_2(Y)$ with $O_2(C_I(t)) \leq O_2(P)$. This is impossible as $O_2(P)$ is abelian.

So take $I^* \in Chev(r)$ for some prime r. As a Sylow 3-subgroup $Q \cap I$ of I is of

3-rank at most 2 and exponent 3, we conclude Q induces a group of inner diagonal automorphisms on I^* and if either $Q \not\leq I$ or 3 divides the order of the center of the universal group of I^*, then $M^* \cong PGL_3^{\epsilon}(q)$, for some $q \equiv \epsilon \bmod 3$. Let $K = \langle Q^{H \cap M} \rangle$; then K is described in 24.3.

If r = 3 then I^* is of characteristic 3-type, so $K^* = Q^*$ and I^* is of GF(3)-type. Hence by Theorem A, $I^* \cong L_3^{\epsilon}(3)$. As $|I^*|$ divides $|G|$ we conclude G is Ru if $I^* \cong L_3(3)$. Moreover an argument above shows $O_2(M) = 1$, so that (5) holds. If G is M_{24} or He then $N_G(T)$ contains no element of order 8 so I^* is not $U_3(3)$ and (6) holds.

Hence we may assume $r \neq 3$. K is described in 24.3 and of course K^* is a homomorphic image of K over a 2-group with $K^* \trianglelefteq C_{M^*}(Z^*)$. Recall also either $Q^* \in Syl_3(I^*)$ and the center of the universal group of I^* is prime to 3, or $M^* \cong PGL_3^{\epsilon}(q)$, $q \equiv \epsilon \bmod 3$. We conclude from 9.1 and 9.2 that one of the following holds:

(a) I^* is defined over GF(2) and $K^* = Q^* \leq I^*$.

(b) G is J_4, $K^* \cong SL_3(4)$, and I^* is defined over GF(2) or GF(4).

(c) $M^* \cong PGL_3^{\epsilon}(q)$, $q \equiv \epsilon \bmod 3$.

In case (a) as $Q^* \in Syl_3(I^*)$, $I^* \cong U_3(3)$ or ${}^2F_4(2)'$. The first case has been handled already. In the second as $|I^*|$ divides $|G|$, G is Ru and $O_2(M) = 1$ by 24.5. That is (5) holds.

In case (b) as $Q^* \in Syl_3(I^*)$, we conclude $I^* \cong G_2(4)$. As $J_2 \leq G_2(4)$, this case is out by an earlier reduction.

Finally in case (c) as each member of $\mathcal{U}_{(H \cap I)^*}(Q^*, 3')$ is an elementary abelian 2-group, $M^* \cong PGL_3(4)$, $PGL_3(7)$, or $PGU_3(5)$. As $|M^*|$ divides $|G|$ either (7) holds or G is Ru and $M^* \cong PGU_3(5)$. Theorem 21.4 supplies a contradiction in this last case.

(24.7) Either

(1) $\mathcal{N}(H) = \{H\}$, or

(2) G is J_4 and $\mathcal{N}(H) = C_G(t)$, where t is the involution in Z(L).

Proof. Let $M \in \mathcal{N}(H)$ and $M^* = M/O_2(M)$. By 24.6 and 24.1, $Z^* \trianglelefteq M^*$. We may assume $O_2(M) \neq 1$, so by 24.5, G is J_4, He, or Ru, and in the last two cases $O_2(M) \cong E_{64}$ and H acts faithfully on $O_2(M)$. But H has no faithful representation over GF(2) of degree 6, so G is J_4. We may assume $M \not\leq C_G(t)$, so $H/\langle t \rangle$ is faithful on $O_2(M) \cap C(t)$. But $C_G(t) = O_2(C_G(t))H$ with H irreducible on $O_2(C_G(t))/\langle t \rangle$. Therefore $C_G(t) \leq M$. Finally $C_G(t)$ is maximal in G (cf. section 30) so the proof is complete.

(24.8) $\langle \mathcal{N}(Q) \rangle = G$.

Proof. Let $\{M\} = \mathcal{N}(H)$. It suffices to exhibit $X < G$ with $Q \leq X \not\leq M$. If $D \in \Delta$ take $X = N_G(D)$. Otherwise by 24.3, G is J_2, where we may take $X \cong U_3(3)$. (cf. a remark below)

In the remainder of this section we verify the tables of maximal overgroups of Q listed in the introduction. The Mathieu group M_{24} is not discussed on the theory that its representation on its Steiner system is the best tool for discussing its overgroups. Thus we appeal to the tables in [19] for M_{24}. If G is Ru or J_4 we leave open the question of the existence and uniqueness of maximal $U_3(3)$-subgroups. Such subgroups seem to be too small to deal with elegantly with the methods of this paper.

If G is J_2 then $\mathcal{U}_G(Q,3') = 1$ by 24.5. Therefore if $M \in \mathcal{N}(Q)$ then by 24.4 and 24.6, either $M = H$ or $L = F^*(M) \cong L_3(4)$ or $U_3(3)$. In the first case M contains a member of Δ, contrary to 24.4. In the second $L = \langle Q,s \rangle$, where s is an element of L of order 4 with $\langle s,u \rangle$ ismorphic to the modular group of order 16, u is of order 8 in $N_L(Q)$, and $\langle s,Z \rangle \cong SL_2(3)$. Let $t = s^2 = u^4$ and $K = C_G(t)$. As t is a square in G, t is 2-central in G. Thus $K \cong A_5/QD$. In particular $D_8 \cong C_K(Z) \leq O_2(K)$ and $v = u^2 \in C_K(Z)$. Therefore $\langle v \rangle$ is uniquely determined by Z and so is $\langle s,Z \rangle = O^2(C_K(v))$. So $L = \langle Q,s \rangle$ is unqiuely determined. Moreover in [26], Hall and Wales use the Brauer trick to show $L_0 = \langle C_G(v),Q \rangle$ is proper in G. Hence $L_0 \cong U_3(3)$, demonstrating the existence of a $U_3(3)$-subgroup of J_2. This completes the treatment of J_2.

Assume G is He. Let $K = N_G(U)$, where U is a 4-subgroup of G with $K/U \cong Z_2/PGL_3(4)$. We may take $Q \leq K$. We saw in 19.9 that if $T \in Syl_2(G)$ then $\mathcal{A}(T) = \{A_1,A_2\}$ with $A_i \cong E_{64}$ and $N_G(A_i)/A_i \cong S_6/Z_3$. In particular we may take $A_i \leq K$. Notice there is $D \in \Delta$ with $N_G(D) \leq K$, so $N_G(D)$ is not maximal in G for $D \in \Delta$. By 24.5, $N_G(A_i)$, $i = 1,2$, and K are, up to conjugation, the unique members of $\mathcal{N}(T)$ with a nontrivial normal $3'$-subgroup.

By 24.6, this leaves only the case where $Q \leq M \cong PGL_3(4)$. If such an M exists, there is X of order 3 in Q with $C_M(X) = X \times J$, where $J \cong A_5$. Then $J \leq E(C_G(X))$, so involutions in J are not 2-central. For each non-2-central involution i in G let U(i) be the root 4-subgroup of G normal in $C_G(i)$. Observe XJ acts on a subgroup V of M isomorphic to E_{16}. Now for $v \in V$, v is fused into J under M and hence is not 2-central. Also v is contained in a 4-subgroup W of V centralized by a subgroup Y of XJ of order 3 and with $W \le [C_G(W),C_G(W)]$. Thus from the structure of $C_G(v)$, $U(v) = W \le V$. Hence V is partitioned by root 4-subgroups. On the other hand G is transitive on pairs of commuting root 4-subgroups and the product of such a pair is conjugate to $A_1 \cap A_2$. Moreover from the structure of $N_G(A_i)$ we see $A_1 \cap A_2$ is not partitioned by root 4-subgroups. Thus G possesses no $PGL_3(4)$-subgroup, so the treatment of He is complete.

Let G be Ru. Then $N_G(Q)/Q \cong SD_{16}$. If $L \cong L_3(3)$ is a subgroup of G containing Q then $L = \langle N_G(D_i) : i = 1,2\rangle$, for $D_i \in \Delta$ with $\tilde{D}_i = C_{\tilde{Q}}(t_i)$ and $E_4 \cong \langle t_1,t_2\rangle = I \le N_G(Q)$. Up to conjugacy in $N_G(Q)$ there is a unique choice for I, so L is determined up to conjugacy in G. Indeed there are just two choices for IQ and hence at most two choices for L. Also there is ${}^2F_4(2) \cong M \le G$. (The stabilizer of a point in the rank 3 representation of Ru.) By 21.4, M contains two $Aut(L_3(3))$-subgroups and E(M) is generated by these subgroups. So M is uniquely determined by Q and neither the normalizer of a $L_3(3)$-subgroup nor that of a member of Δ is maximal in G.

Assume $E_{64} \cong V \in \mathcal{U}_G(Q,2)$. Then by 24.5, $V = [V,Z]$ and hence if $Z \ne X$ is of order 3 in Q then $C_V(X) = U$ is a 4-subgroup of V. Let $t \in U^{\#}$, $u \in U - \langle t\rangle$, $K = C_G(t)$, and $P = O_2(K)$. As 3 divides $|K|$, t is 2-central. From 12.12, $C_P(X) \cong Q_8$ and $K/P \cong S_5$. In particular $C_P(X)$ contains no involutions so $u \notin P$ and then as u centralizes X, u induces an outer automorphism on $E(K/P) \cong A_5$. $C_K(X)$ has semidihedral Sylow 2-subgroups so U is determined up to conjugacy in $C_K(X)$. Let $E = \Phi(P)$ and $\bar{K} = K/\langle t\rangle$. Then from 12.12, $m(C_{\bar{E}}(u)) = 2 = m(\bar{E})/2$ and $m(C_{P/E}(u)) = 3 = m(P/E)/2$. Therefore $|C_P(u)| = 32$ and $C_K(u)/C_P(u) \cong C_{K/P}(u) \cong Z_2 \times S_3$. In particular $V = [C_P(u),X]U = O_2(C_K(U))$ is determined by U, and V is determined up to conjugacy in G.

Conversely let U be a 4-subgroup of $C_K(X)$ and $V = [C_P(U),X]U$. From 24.1 each 4-subgroup of $C_G(X)$ is invariant under a 9-group, so we may take $U = [U,Z]$. V is Sylow in $C_G(U)$ so by a Frattini argument we may take $XZ \le N_G(V)$. Then $V = \langle C_V(y) : y \in XZ^{\#}\rangle$ and $C_V(y) \cong E_4$ or 1. Also $U \not\le \Phi(V)$ and Z is irreducible on U, so $V = U \times [V,X]$. So as $|V| = 64$, we conclude $V \cong E_{64}$. By 24.5, $V \in \mathcal{U}^*_G(Q,3')$ and we showed in the last paragraph that V is determined up to conjugacy in G.

Next from the action of u on P, $[E,V] = E \cap V$ so $E \le N_G(V) = M$. Also $P/E = (P_2/E) \times (C_P(X)E/E)$ where $P_2 = [P,X]$. $VE/E = [P_2/E,u]$, so P_2 acts on VE. Thus P_2 acts on $J(VE) =$

V. Finally an element of order 4 in $C_P(X)$ inverted by u acts on V as does $C_K(U)$. Hence $|N_K(V)| = 2^{13}3$. We calculate that $N_K(V)$ has orbits {t}, $(V\cap E) - \langle t\rangle$, $(V\cap P) - E$, and $V - P$ of order 1,6,24, and 32, respectively. Further Z fuses t into $V - P$ and $|t^M|$ divides the order of $L_6(2)$. We conclude $t^M = V^{\#}$ and $|M/V| = |V^{\#}||N_K(V):V| = |G_2(2)|$. Therefore by 24.6, $M/V \cong G_2(2)$.

We've also seen that if U is a 4-group centralizing X of order 3, then $V = O_2(C_G(U)) \cong E_{64}$. In particular V is not Sylow in $L_3(4)$, so there is no $PGL_3(4)$ subgroup of G. This completes the analysis of the case $G \cong Ru$.

So we are reduced to $G \cong J_4$. Looking ahead to 30.2 we find G has maximal subgroups K and L where $K \cong Aut(M_{22})/Z_3/D^6$ is the centralizer of a 2-central involution t and $L = N_G(V)$ where $V = J(S_0) \cong E_{2^{11}}$ for $S_0 \in Syl_2(G)$, and $L/V \cong M_{24}$. We may assume $Q \leq K\cap L$. We will show any 2-local containing Q and any overgroup of Q isomorphic to M_{24} or M_{12} is contained in a conjugate of K or L. As we are avoiding consideration of $PGL_3(4)$ and $U_3(3)$ subgroups, this will complete our analysis, given 24.6.

Let $E_9 \cong D \leq Q$. By 24.4, $D \in \Delta$. There are two conjugates L and L^r, $r \in N_G(Q)$, of L containing Q and $N_G(D)$ is contained in one of these, say L. Moreover there is a unique 4-group E in $Out_G(Q)$ stabilizing D and the other maximal subgroup of Q invariant under E is conjugate to D in L.

Now if $M_{12} \cong M \leq G$ then $M = \langle M_1,M_2\rangle$, where $M_i \cong SL_2(3)/E_9$ and $O_3(M_i) = D_i$ is determined by a 4-subgroup E_0 of $Aut_G(Q)$ acting on D_1 and D_2. There are two conjugate choices for E_0 in $Out_G(Q) \cong SD_{16}$, so without loss $D_1 = D$ and $D_2 = D^s$, for suitable $s \in N_L(Q)$. Therefore $M \leq \langle N_G(D),s\rangle \leq L$.

Similarly suppose $M_{24} \cong M$ is an overgroup of Q. This time up to conjugation M $= \langle M_1,N_M(X)\rangle$ with $SL_2(3)/E_9 \cong M_1 \leq N_G(D)$, $N_M(X) \cong S_3 \times L_3(2)$, and $XZ \neq D$ or D^s, since all subgroups of D and D^s of order 3 are 3-central in M. In particular there is an involution k in $N_M(Q)$ inverting X and centralizing $E(C_M(X)) = Y$. Then k induces an outer automorphism on $E(C_G(X)) = A$ centralizing Y, so $C_A(k)^{\infty} \cong L_3(2)/E_8$. Now $\langle k\rangle = \Phi(S)$, where $D_8 \cong S \in Syl_2(N_M(Q))$, so $S\langle t\rangle$ and $\langle k\rangle = \Phi(S\langle t\rangle)$ is determined in $N_G(Q)$ modulo Q. Further $C_L(X) \cong L_3(2)/E_8$ is centralized by $\Phi(S\langle t\rangle)$, so $C_G(X\langle k\rangle) \leq L$. Therefore $M = \langle M_1,C_M(X\langle k\rangle)\rangle \leq L$.

Finally assume M is a 2-local overgroup of Q and let $S \in Syl_2(M)$. If S is Sylow in G then by section 30, M is contained in K, L, or L^r. So assume otherwise. We will show there is a characteristic subgroup C of S normal in M, contradicting S not Sylow in G. Indeed C will be $\Omega_1(Z(S)) = I$ or $J(S)$. So assume neither I nor $J = J(S)$ is normal in M.

Let $R = O_2(M)$ and $W = \Omega_1(Z(R))$. By 24.6, $M^* = M/R \cong M_{24}$ or an extension of M_{12}, $U_3(3)$, $PGL_3(4)$, M_{22}/Z_3, $SL_3(4)$, A_6/Z_3, A_7/Z_3, 3^{1+2}, or $SL_2(3)/E_9$.

Claim $R = C_M(W)$. For if not either $I \trianglelefteq M$ or $C_M(W) \le SQ$, and by assumption it is the latter. However there is a unique maximal member V_0 of $\mathcal{U}_{C(Z)}(Q,2)$ and $V_0 \le V$. So $V \le C_M(W)$ and by a Frattini argument, $M = C_M(W)N_M(V) \le SQN_M(V) = N_M(V) \le L$, contrary to the choice of M.

So $R = C_M(V)$. As J(S) is not normal in M, $J(S) \not\le R$. This says there is $A \in \mathcal{A}(S)$ with $A^* \ne 1$ and $m(A^*) \ge m(W/C_W(A))$. By the representation theory of M^* this cannot happen unless M^* is $G_2(2)$ or an extension of $SL_3(4)$ or A_6/Z_3. For example this is proved if $M^* \cong M_{24}$ in 31.9. The other cases are discussed in [4]. The case $F^*(M^*) \cong U_3(3)$ will be done here anyway.

Suppose $B^* = F^*(M^*) \cong SL_3(4)$ or A_6/Z_3. The existence of A^* forces $m(W) = 6$ or $B^* \cong SL_3(4)$ and W the sum of two equivalent natural modules for B^*. In the latter case, or if $B^* \cong A_6/Z_3$, we have $J(O_2(C_M(i))) = J(R)$ for $i \in I^\#$. Thus $P = O_2(C_G(i)) \le O_2(C_M(i))$, so $\langle i\rangle = C_G(P) = C_W(O_2(C_M(i)))$. We conclude $BP^* \cong S_6/Z_3$ and $P^* \cong E_8$; for example if $W = W_1 \oplus W_2$ is the sum of 2 modules for $SL_3(4)$, then $C_W((O_2(C_M(i))) = C_{W_1}(O_2(C_M(i))) \oplus C_{W_2}(O_2(C_M(i)))$ is of order 4. As $[W,P] \not\le \langle i\rangle$ also $W \not\le P$. But from the representation of P on W, $m(\langle i,[P,W]\rangle) \le 3$ for each $w \in W$. This contradicts $m([j,P/\langle i\rangle]) \ge 3$ for each involution j in $C_G(i)/P$.

So $m(W) = 6$ and $B^* \cong SL_3(4)$. Now up to conjugation $C_G(Z)$ has a unique subgroup B_0 possessing a $SL_3(4)$-homorphic image. Further $Z(B_0) \cong Z_6$, so $C_R(Q) \ne 1$. This will lead to a contradiction below.

So assume $B^* = F^*(M^*) \cong U_3(3)$. Then $m(S^*) = k = 2$ or 3 for $M^* \cong U_3(3)$ or $G_2(2)$, respectively. Notice for each noncentral 2-chief section U of R, $m([U,Z]) \ge 6$ and $m([U,X]) \ge 4$ for X non-3-central of order 3. But if a^* is an involution in $M^* - B^*$ then a inverts a conjugate of Z^* so $m([U,a]) \ge 3$. Similarly if b^* is an involution in B^* then $m([U,b]) \ge 2$. As $C_B*(b)$ is maximal in B^*, $C_U(b) \ne C_U(b_1)$ for a conjugate $b_1^* \ne b^*$. Applying these remarks to the action of A^* on a section U of W we conclude $M^* \cong G_2(2)$, $m(A^*) = m(W/C_W(A)) = 3$, $m([W,a]) = 3$, and $m([W,b]) = 2$. Then M^* is generated by two conjugates of A^* so $m(W/C_W(M)) \le 2m(W/C_W(A)) = 6$. So $[W,M] = [W,Z]$ is of rank 6 and $W = [W,M] \oplus C_W(M)$.

It's now easy to see M is transitive on $[W,M]^{\#}$ and for $w \in [W,M]^{\#}$, $C_M(w)$ acts on a hyperplane W_w of [W,M] containing w. Then $w \mapsto W_w$ is a symplectic polarity on [W,M] preserved by M^*, so we may regard [W,M] as a 6-dimensional symplectic space over GF(2) for M^*.

We have reduced to [W,M] of rank 6 and either $M^* \cong G_2(2)$ or $B^* \cong SL_3(4)$ with $C_R(Q) \neq 1$ in the latter case.

Let W_0 be the intersection of the groups $O_2(C_G(w))$ as w ranges over $[W,M]^{\#}$. If $[W,M] \leq W_0$ then by 17.6, $Aut_G([W,M]) = L_6(2)$, a contradiction. So $[W,M] \not\leq O_2(C_G(w))$ and as M is irreducible on [W,M], $W_0 \cap [W,M] = 1$.

Suppose $C_R(Q) \neq 1$. Then $C_R(Q) = \langle t\rangle$. As Q is irreducible on [W,M] and $m(K/O_2(K)) = 5$, $[W,M] \leq O_2(K)$. But then $t \in W_0 \trianglelefteq M$ so as $W_0 \cap [W,M] = 1$, $1 \neq C_W(M)$. That is $\langle t\rangle = C_W(M)$. But then $M \leq K$, contrary to hypothesis.

Therefore $C_R(Q) = 1$. Thus $W = [W,M]$ and $M^* \cong G_2(2)$. Also Q is faithful on each 2-chief factor U of M, and hence by a remark above, $m(U) \geq 6$. Hence as $|G|_2 = 2^{21}$ and $|M^*|_2 = 64$, we conclude either R = W or R is special with center W and $6 \leq m(R/W) \leq 9$. Moreover in the latter case as $C_R(Q) = 1$ and $N_M(Q)$ is irreducible on Q/Z, either $[R/W,Z] = R/W$ is of rank 6 or $m(R/W) \geq 10$, (since 4 is the minimal dimension of a Q/Z-faithful $N_M(Q)/Z$-module) so actually $m(R/W) = 6$.

Now $I = \Omega_1(Z(S))$ is a point of W and $C_M(I)/R \cong D_{12}/(Z_4)^2$. Let $P = O_2(C_G(I))$. Then $(P\cap M)^* \cong E_8$ or E_4. In the later case $J(R) = J(R(P\cap M)) \trianglelefteq N_G(R(P\cap M))$, so $P \leq M$. Then as $m(C_G(I)/P,P/I) > 2$ (cf. 19.2) and $m(PR/R) \leq 2$, $W \leq P$, contrary to an earlier remark. Thus $(P\cap M)^* = \Omega_1(O_2(C_M(I)^*) \cong E_8$. In particular there exists $a \in P\cap M - B$. Also there is Y of order 3 in M with $C_W(Y) = \langle I,w\rangle$, w not in the unique $C_M(I)$-invariant hyperplane W_I of W, and a inverting Y^*. As $W \not\leq P$ it follows that $w \notin P$. Of course $[a,w] = i$ generates I as $m([W,a]) = 3$, $C_W(Y) = \langle I,W\rangle$, and $m([Y,W,a]) = 2$. So by lemma 2.6, $[w,X] \leq P$, where X is of order 3 in $C_G(I)$ with $XP \trianglelefteq C_G(I)$. Observe X induces a GF(4)-vectorspace structure on P/I preserved by $C_G(XP/P)$.

Suppose R = W. Let S_0 be a Sylow 2-subgroup of G containing S, $U = Z_2(S_0)$ and U_0/I the 1-dimensional GF(4)-subspace of P/I containing U/I. Then $U \leq W$ and either $U_0 \leq W$ or $[U_0,W] = U$. In the latter case w centralizes U but not U_0/I, contrary to the previous paragraph. Thus $U_0 \leq W$. But now $U_0 = \langle U^{C_M(I)}\rangle \trianglelefteq C_M(I)$ and $N_M(U)^*$ is the second parabolic of M^* containing S^*, so $M = \langle N_M(U),N_M(U_0)\rangle$. On the other hand looking ahead to section 30 and the structure of the 2-parabolics of G, $N_G(U) \leq$

$N_G(U_0)$ and $N_G(U_0)$ has no $U_3(3)$-section, a contradiction.

So R is special with $W = Z(R)$ and $m(R/W) = 6$. Indeed $R/W = [R/W,a]\oplus[R/W,a^z]$, since we can take a to invert Z^*. Therefore $R = (P\cap R)(P^z\cap R)$. Also $[P\cap R,P^z\cap R] \leq P\cap P^z\cap W = 1$, so $\Phi(R) = \Phi(P\cap R)\Phi(P^z\cap R) = II^z$. This is a contradiction as R is special with $\Phi(R) = W$.

25. Overgroups of elementary abelian p-groups.

In this section p is a prime, G is a finite group, and A is a noncyclic elementary abelian p-subgroup of G such that $\mathcal{N}(A) = \{M\}$. Using a Frattini argument it's easy to show:

(25.1) $\ker_M(G)/O_p(\ker_M(G))$ is nilpotent.

So in the remainder of this section take $\ker_M(G) = 1$.

(25.2) If $1 \neq H \trianglelefteq G$ then $G = HA$.

Proof. As $\ker_M(G) = 1$, $H \not\leq M$, so $HA \not\leq M$.

(25.3) $F^*(G)$ is a nonabelian simple group and $G = AF^*(G)$.

Proof. Let H be a minimal normal subgroup of G. By 25.2, G = HA. Thus either

(i) H is an elementary abelian r-group for some prime r and A is irreducible on H,

or

(ii) H is the direct product of the A-conjugates of some simple component L of G.

$C_A(H) \trianglelefteq G$, so as $\ker_M(G) = 1$, $C_A(H) = 1$. Thus if (i) holds then A is faithful and irreducible on H. This is impossible as A is noncyclic. So (ii) holds. If H is simple we are done, so assume not and let $B = N_A(L)$. Then each $a \in A - B$ is semiregular on L^A, so $C_H(a)$ is isomorphic to the direct product of n/p copies of L, where $n = |L^A|$. In particular L is the projection of $C_H(a)$ on L. On the other hand if $b \in B^{\#}$, then by a result of Thompson, $1 \neq C_L(b)$. Therefore as $C_L(b) \leq M$, $M \cap L \neq$ if $B \neq 1$. Also if $B = 1$ then for each $x \in L^{\#}$, $\langle x,A\rangle \neq G$, so $x \in M$. Thus in any case $M \cap L \neq 1$. Therefore as $C_H(a) \leq M$ and L is the projection of $C_H(a)$ on L, $L \leq M$. But now $H = \langle L^A\rangle \leq M$, a contradiction.

(25.4) If H is a finite group with $F^*(H)$ a sporadic simple group and T is a noncyclic Sylow p-subgroup of H, then $H = \langle \mathcal{N}(T) \rangle$.

Proof. Let $L = F^*(H)$. If L is a Mathieu group appeal to the tables of maximal subgroups in [19]. If p = 2 and L is one of the groups of section 12, then appeal to the results of section 12. If T is abelian appeal to section 16. (The results are summarized in the tables in the introduction.) If p = 3 and L is Ly or Sz appeal to 11.7. If p = 3 and L is J_3 or F_3, appeal to sections 13 and 14. The groups in section 24 are dealt with in 24.8.

Thus we are reduced to the case where H possesses a large symplectic p-subgroup Q. If Q is of width 1, appeal to section 21. In the remaining cases 20.3 reduces us to p = 2 and $L \cong Co_2$ or $Q \cong 3^{1+4}$ and $G \cong Co_1$, Co_2, Co_3, or Mc. Here appeal to 20.2 together with 19.3.6, 19.6, 19.10, or the existence of a $M_{22}/2^{10}$ subgroup of Co_2. This last can be seen from the action of Co_1 on the Leech lattice.

(25.5) One of the following holds:

(1) $G \in Chev(p)$ and G is of Lie rank 1, or $G = {}^2G_2(3) = Aut(L_2(8))$.

(2) $G \cong A_{2p}$, $p > 3$, and $A \cong E_{p^2}$.

(3) $G \cong Aut(Sz(32))$ and $A \cong E_{25}$.

Proof. By 25.3, $L = F^*(G)$ is a nonabelian simple group and if $A \le L$ then G = L. If L is an alternating group then (2) holds by 7.5, while by 25.4, L is not a sporadic group. So $L \in Chev(r)$ for some prime r.

If r = p then either L is of Lie rank 1 or p = 2 and L is of Lie rank 2, or A acts on distinct parabolics of L generating L. The last case is certainly out, while the second is easily handled by an argument similar to one to follow. (cf page 421 in [3].) So we can assume L is of Lie rank 1 but there is some $a \in A - L$. But now unless $G = {}^2G_2(3)$, $C_L(a)$ is not contained in a Borel group stabilized by A, a contradiction.

So $r \neq p$. By Theorem B in [3] we may assume p is odd. As A is abelian,

$\Gamma_{1,A}(G) \le M \ne G$, so G is described in Theorem 1 or 2 of [34]. In particular $r = 2$.

Suppose A does not induce a group of inner diagonal automorphisms on L. Then G is described in Theorem 2 of [34]. Notice Aut(Sz(32)) and $Aut(L_2(8)) \cong {}^2G_2(3)$ appear on our list of exceptions, so we need not consider these cases. Without loss A is a maximal elementary abelian p-subgroup of G, so $A \cap L \ne 1$. Thus if some $a \in A^{\#}$ induces a field automorphism on L, then so does another $b \in a(A \cap L)$ by Lang's Theorem. However by Theorem 1 in [34], $C_L(a)$ and $C_L(b)$ are maximal in L, so $L = \langle C_L(a), C_L(b)\rangle \le M$, a contradiction.

Thus we are left with case (iv) of Theorem 2 of [34] and G is not $O^{3'}(Aut(D_4(8))$. Then $p = 3$ and A contains an element z of order 3 3-central in G. Let $A \le T \in Syl_3(C_G(z))$. If $G \cong Z_3/\Omega_8^+(2)$ then $G = \langle \mathcal{N}(A)\rangle$ by 20.3, a contradiction. If $G \cong Z_3/\Omega_8^+(4)$ then $G = \langle C_G(\alpha), C_G(z)\rangle$, where α is a field automorphism of L of order 2 centralizing T, again a contradiction. Finally if $G \cong Aut({}^3D_4(2))$ then by 21.6, $N_G(\langle z\rangle)$ is maximal in G, so as $C_G(a) \not\le N_G(\langle z\rangle)$ for $a \in A - \langle z\rangle$, we again have a contradiction.

Therefore we may take A to induce a group of inner-diagonal automorphisms on L. Hence G is described in Theorem 1 of [34]. $G_2(2)' \cong U_3(3)$ and $Sp_4(2)' \cong L_2(9)$, so we can ignore these cases. Again we may take A to be a maximal elementary abelian p-subgroup of G, $z \in A^{\#}$ with z p-central in G, and $A \le T \in Syl_3(C_G(z))$. If $L \cong F_4(2)$, ${}^2E_6(2)$, or ${}^2F_4(2)'$ then appeal to 20.3 or 16.12.2 to get $\langle \mathcal{N}(T)\rangle = G$. If $L \cong F_4(4)$ or $E_8(4)$ then T centralizes a field automorphism α of order 2 and by Theorem 1 in [34], $C_G(\alpha)$ is maximal in G. Thus as $C_G(z) \not\le C_G(\alpha)$, $G = \langle C_G(\alpha), C_G(z)\rangle \le M$, a contradiction. Similarly if $L = E_6(4)$ or $E_6(2)$ then T centralizes a graph-field automorphism α and we obtain the same contradiction. Thus if L is exceptional then $G = E_7(2)$ or $E_8(2)$ and $p = 3$. From the discussion on page 391 of [34] we may choose z so that $C_G(z)$ has a component ${}^2E_6(2)/Z_3$. Further T is contained in the normalizer N of a maximal torus $N_0 \cong E_{3^k}$ and N/N_0 is isomorphic to the Weyl group of E_k. Now $G = \langle C_G(z), N\rangle \le M$, a contradiction.

So take L to be a classical group described in case (i) of Theorem 1 of [34]. Adopt the notation of section 4 of [34]. In particular L is the central factor group of G_0 = SL(V), Sp(V), SU(V), or Ω(V), for a suitable vector space V over F = GF(4) if L $\cong$ $U_n(2)$ or $L_n(4)$ and F = GF(2) otherwise. Moreover E_0 is the preimage of A in GL(V). If E_0 is abelian then by 3.2 and 4.1 in [34], either there exist A-invariant decompositions $V = V_{i1}+V_{i2}$, i = 1,2, with $G_i = N_{G_0}(\{V_{i1},V_{i2}\}) \neq G_0$ and $G_0 = \langle G_1,G_2\rangle$ or $G \cong \Omega_8^+(2)$ or $U_4(2)$. In the first case we have a contradiction. The second can be treated just as $Z_3/\Omega_8^+(2)$ was. In the third $U_4(2) \cong PSp_4(3)$, so we are O.K.

Therefore we may take E_0 nonabelian. Hence L $\cong$ $U_n(2)$ or $L_n(4)$, n $\equiv$0mod3. Let $Z = Z(G_0)$ and $Z \leq E \leq E_0$ with $E \cong E_9$. Then $V = \bigoplus_{D\in\Gamma(E)} C_V(D)$, where $\Gamma(E)$ is the set of subgroups of E of order 3 distinct from Z with $C_V(D) \neq 0$.. Moreover if L is unitary this sum is orthogonal. If $E \leq Z(E_0)$ then $E_0 \leq N(C_V(D))$ for each $D \in \Gamma(E)$. Also $N(C_V(D))$ is a maximal parabolic if L $\cong$ $L_n(2)$ while $N(\{C_V(D),C_V(D)^{\perp}\})$ is maximal in G_0E_0 if L is unitary by 6.8 or 6.9. So $E \not\leq Z(E_0)$. Then E_0 acts transitively on $P = \{C_V(D) : D \in \Gamma(E)\}$ and applying 6.9 to the groups in case (i) of Theorem 1 of [34], either N(P) is maximal in G_0E_0 or L $\cong$ $L_3(4)$ or $U_6(2)$. $U_6(2)$ is handled by 23.1 and $L_3(4)$ is generated by A_6-subgroups over A. Thus $\mathcal{N}(A) = \{N(P)\}$, so E is the unique E_9-subgroup of E_0. We conclude A $\cong$ E_9 and E_0 is the nonabelian group of order 27 and exponent 9. As $\dim(C_V(D)) > 1$ we can write $C_V(D) = U_1+U_2$, with this decomposition orthogonal if L is unitary. Then $V_i = \langle U_i^{E_0}\rangle$ is an E-invariant subspace with $G_0E_0 = \langle N(V_1),N(P)\rangle$.

This completes the proof of 25.5.

26. Small overgroups of large symplectic p-subgroups.

In this section assume Q is a large symplectic p-subgroup of the finite group G with Q of width at least 2, $L = F^*(G)$ is simple, $G = LQ$, and L is not of Lie type and characteristic p. Exclude case (2) of Theorem A. Let $Z = \langle z \rangle = \Omega_1(Z(Q))$, $H = N_G(Z)$, and $T \in Syl_p(H)$.

(26.1) Let $X = \langle Q^X \rangle \leq G$ with $V = \langle Z^X \rangle$ abelian and $V \not\leq Q$. Then

(1) If $|Q:C_Q(V)| = p$ or $m(Q \cap V) \leq 2$ then $p = 3$, $G \cong Co_1$, VQ/Q induces a transvection on Q/Z, $m(V) = 3$, and $X/C_X(V) \cong \Omega_3(3)$.

(2) If $|Q:C_Q(V)| = p^2$ or $m(Q \cap V) = 3$ then $m([Q/Z,V]) = 2$, and either $p = 2$ or $Q \cong 3^{1+4}$.

Proof. Assume the hypotheses of (1). Then by 17.6.5 and 17.6.6, $m(V) = 3$ and $X/C_X(V) \cong \Omega_3(p)$. In particular $|Q:C_Q(V)| = p$ so V induces a group of transvections on Q/Z. Therefore 19.2 completes the proof.

Next assume the hypotheses of (2). Then $m([Q/Z,V]) = 2$ so by 19.2 either $p = 2$ or $Q \cong 3^{1+4}$.

(26.2) Let $Q \cong 3^{1+4}$ and $Q \leq M \leq G$ with $T \cap M \in Syl_3(M)$, $M \not\leq H$ and $O_3(M) \neq 1$. Then one of the following holds:

(1) $M \leq N_G(J(T))$.

(2) $G \cong Co_1$ and $M \leq N_G(V)$ where $V = C_G(C_Q(Z_2(T))) \cong E_{27}$, $R = O_3(N_G(V))$ is special of order 3^7 with $V = Z(R)$, and $N_G(V)/R \cong GL_2(3)*GL_2(3)$ with $O^2(N_G(V))/R$ acting naturally as $\Omega_4^+(3)$ on R/V and $Aut_G(V) \cong O_3(3) \cong Z_2 \times S_4$.

(3) $G \cong F_4(2)$ and $M = N_G(V_j)$ for $j = 1$ or 3, where $E_{27} \cong V_j \leq Q$, $\langle N_G(V_1),N_G(V_3)\rangle = N \cong L_4(3)$, and $N_G(V_j)$ is a maximal parabolic of N.

Proof. By 17.6 $V = \langle Z^G \rangle$ is abelian and $F^*(M) = R = O_3(M)$. Suppose first that Z is not weakly closed in Q. Then by 19.5 and 19.6, $G \cong Z_3/\Omega_8^+(2)$, $F_4(2)$, or F_5, $J = J(T) \cong E_{81}$, and J is the natural module for $O^{3'}(N_G(J))/J \cong \Omega_4^+(3)$. By 19.8, $\mathcal{Y}(T) = (Z_1)^Q \cup (Z_3)^Q$ is of order 6 for some $Z_i = Z^{g_i}$. If $V = ZZ^g$ for some $Z^g \in \mathcal{Y}(T)$ then $T = Q(Q^g \cap H)$, so $N_H(V) \leq N_H(T) \leq N_G(T)$, and therefore $N_G(V) = \langle Q^g, N_H(V)\rangle \leq N_G(J)$. If $V \leq Q$ then $m(V) = 2$ or 3 and by 17.6, $O^{3'}(Aut_G(V)) = SL(V)$. We have handled the case $m(V) = 2$ so take $m(V) = 3$.

If $G \cong F_4(2)$ the discussion in section 23 shows $V = Q \cap Q^{g_j}$ for $j = 1$ or 3, and that (3) holds. If G is $Z_3/\Omega_8^+(2)$ then $T \not\leq L = G^\infty$ whereas $N_G(V)^\infty$ contains a Sylow 3-group of G. Finally if G is F_5 then $H/Q \cong Z_4{*}SL_2(5)$, so H has no $GL_2(3)$ section, whereas $N_H(V)$ has such a section.

So $V \not\leq Q$. By 26.1, $m(V \cap Q) \geq 3$, so $m(V) \geq 4$. Therefore $V = J$, completing the proof in this case.

So we may assume Z is weakly closed in Q. Therefore G is a Conway group, Mc, $U_6(2)$, or $U_4(3)$ by 19.6. Also $V \not\leq Q$ so by 26.1 either $m(Q \cap V) \geq 3$ or G is Co_1 and V satisfies the conclusion of 26.1.1. In the first case $m(V) \geq 4$. But unless G is Co_1 or Co_3, J is the unique abelian subgroup of T of rank 4, so $V = J$ and we are done. On the other hand if G is Co_1 or Co_3 then by 23.6, $N_G(V) \leq N_G(J)$.

Thus we may take G to be Co_1 and assume V satisfies the conclusion of 26.1.1. Then $|Q:C_Q(V)| = 3$ so $V = C_G(C_Q(V \cap Q))$ and $V \cap Q = Z_2(T)$. Also from the structure of H, $C_G(V)/R \cong SL_2(3)$. It is now easy to see that (2) holds.

Now assume G is Co_1 and let's see that $C_G(C_Q(Z_2(T)))$ does satisfy the conclusion of 26.2.2. Namely $N_G(J)/J \cong M_{12}/Z_2$ has a subgroup $X/J \cong GL_2(3)/E_9/Z_2$ with $T \leq X \not\leq H$, so $\langle Z^X\rangle = V_0$ satisfies the hypothesis of V, and hence $V_0 = C_G(C_Q(Z_2(T)))$.

Define $\mathcal{Z}(T)$ consist of those conjugates Z^g of Z such that $Q^g \cap H \leq T$ and for $X = \langle Q, Q^g\rangle$ and $V = \langle Z^X\rangle$, V is the natural module for $X/C_X(V) \cong \Omega_4^-(p)$.

(26.3) Let $X \in \mathfrak{D}(G,T,Q)$ with $X \not\leq H$ and $O_p(X) \neq 1$. Then $X = \langle Q, Q^g\rangle$ and one of the following holds:

(1) $Z^g \in \mathcal{Y}(T)$.

(2) $p \leq 3$, $m(H/Q, Q/Z) \leq 2$, and $Z^g \in \mathcal{Z}(T)$.

(3) $p = 3$, $G \cong Co_1$, and $\langle Z^X\rangle = C_G(C_Q(Z_2(T)))$.

Proof. As $X \in \mathfrak{D}(G,T,Q)$, $\mathcal{N}_X(Q) = \{Y\}$. Let $K = \ker_Y(X)$ and $X^* = X/K$. Let $V = \langle Z^X\rangle$ and $\bar{X}$

$= X/C_X(V)$. By 17.6, X is irreducible on V so $O_p(\bar{X}) = 1$. By 25.1, (which holds even when $\bar{Q}$ is cyclic) $\bar{K}$ is nilpotent so $\bar{K}$ is a p'-group and hence $\bar{Q} \cong Q^*$.

If Q^* is cyclic then by 26.1, 26.2, and 17.6.3 either (1) or (3) holds. So assume Q^* is noncyclic.

Claim $\bar{K} = Z(\bar{X})$. If not $[Q,\bar{K}] \neq 1$, so there is $M \in \mathfrak{D}(QK,T,Q)$ with $M \not\leq H$. QK, and hence also M, is solvable. Thus by 25.5, $Q\ker_N(M)/\ker_N(M)$ is cyclic, where $\{N\} = \mathcal{N}_M(Q)$. By the last paragraph it follows that $M/O_p(M) \cong SL_2(p)$ and M satifies (1) or (3). In particular as M is solvable $p \leq 3$ and $\bar{K}$ is a $\{2,3\}$-group. If $\mathcal{Y}(T) = U^Q$ is of order p then $\langle Z^M \rangle = ZU$ and $Q \cap O_p(M) = C_Q(U)$ is independent of M and we conclude $C_Q(U)$ centralizes $\bar{K}$. But then from 25.5, $X = \langle (C_Q(U)^X \rangle$ centralizes $\bar{K}$ as desired. Similarly if G is Co_1 then $Q \cap O_p(M) = C_Q(Z_2(T))$ is independent of M and the same argument works. This leaves the case where $\mathcal{Y}(T)$ has order greater than p. Hence G is described in 19.8. If $p = 3$ then 26.2 says (1) holds. So take $p = 2$. Then by 19.8, G is M_{22}, M_{24}, or He. In particular $m_3(G) = 2$, so $m_3(\bar{K}) = 2$. But then $X = X^\infty$ centralizes $\bar{K}$.

Therefore $\bar{K} = Z(\bar{X})$. Also if $m(Q^*) = 2$ then $p \leq 3$ by 26.1, so by 25.5, $\bar{X}$ is quasisimple with $\bar{X}/Z(\bar{X}) \cong L_2(p^e)$, $U_3(p^e)$, $Sz(2^e)$, or $^2G_2(3^e)$. Next $Q \leq T \cap X = S \in Syl_p(X)$, $\bar{Q}$ is weakly closed in $\bar{S}$, and $\Phi(\bar{Q}) = 1$. It follows that $\bar{Q} = \bar{S}$ if $\bar{X}/Z(\bar{X}) \cong L_2(p^e)$, $\bar{Q} = Z(\bar{S})$ if $\bar{X}/Z(\bar{X}) \cong U_3(p^e)$ or $Sz(2^e)$, and $Z(\bar{S}) \leq \bar{Q}$ if $\bar{X}/Z(\bar{X}) \cong {}^2G_2(3^e)$. But then if $\bar{X}/Z(\bar{X})$ is unitary, $\langle \bar{Q}, \bar{Q}^x \rangle \cong SL_2(p^e)$ for some $x \in X - H$, contradicting $\{Y\} = \mathcal{N}_X(Q)$.

Next $C_Q(V) \leq C_Q(V \cap Q)$ and $m(Q/C_Q(V \cap Q)) = m(V \cap Q) - 1$, so $f = m(\bar{Q}) \geq m(V \cap Q) - 1$. Also $[Q,V] \leq V \cap Q$, so $m([Q,V]) \leq f+1$. On the other hand $X = \langle Q, Q^x \rangle$ for $x \in X - H$, so $m(V) \leq 2m([Q,V]) \leq 2(f+1)$. However if $\bar{X}/Z(\bar{X}) \cong Sz(2^e)$ or $^2G_2(3^e)$ then $m(V) \geq 4e$ or 6e, respectively, as $\bar{X}$ contains an element of prime order dividing $2^{4e}-1$ or $3^{6e}-1$ but not p^i-1 for $i < 4e$ or 6e, respectively. Therefore $f \geq 2e-1$ or $3e-1$, respectively, so as $|\bar{S}| = 2^{2e}$ or 3^{3e}, respectively, and as $\bar{Q}$ is weakly closed in $\bar{S}$, it follows that either $\bar{Q} = \bar{S}$ or $\bar{X} \cong {}^2G_2(3)$. The former as out as $\bar{S}$ is nonabelian. In the latter as $m(V) \geq 6$, we conclude $G \cong Co_1$ and $V = J(T)$. But by 19.3, $N_G(J(T))$ has no $^2G_2(3)$-section.

This leaves $\bar{X}/Z(\bar{X}) \cong L_2(p^e)$ and $f = e$. We have shown $m([V,Q]) \leq e+1$. Of course $Z = C_V(Q)$. Hence the action of X on V is described in 4.1. If $V \leq Q$ then (1) holds by 17.6.3. By 26.1 we may take $m(V) \geq 4$, so by 4.3, $m(V) = 4$ and V is the natural module for $\bar{X} \cong \Omega_4^-(p)$. In particular $e = 2$, so (2) holds by 26.1.

(26.4) Assume $\mathcal{Z}(T)$ is nonempty and let $z^g \in \mathcal{Z}(T)$, $X = \langle Q,Q^g\rangle$, and $V = \langle z^X\rangle$. Then one of the following holds:

(1) $Q \cong 3^{1+4}$, $G \cong Co_2$, Mc, or $U_6(2)$, Q is transitive on $\mathcal{Z}(T)$, and $V = J(T)$.

(2) $Q \cong 3^{1+4}$, $G \cong Co_1$ or Co_3, $V \le J(T)$, and $N_G(V) \le N_G(J(T))$.

(3) $p = 2$, $G \cong Co_2$ or $L_4(3)$, and Q is transitive on $\mathcal{Z}(T)$.

(4) $p = 2$, $G \cong M_{22}$, Sz, or Co_1, T is transitive on $\mathcal{Z}(T) = V^{\#}$, $O^2(Aut_G(V)) \cong A_6$, and $\langle Q^g,T\rangle = \langle Q^h,T\rangle$ for some $z^h \in \mathcal{Y}(T)\cap V$.

(5) $p = 2$, $G \cong U_4(3)$, and $\mathcal{Z}(T) = (V - Q) \cup (V - Q)^{\alpha}$ for some outer automorphism α of G.

Proof. Let $H^* = H/Q$ and $\tilde{H} = H/Z$. As $\mathcal{Z}(T)$ is nonempty, 26.1 says $m(H^*,\tilde{Q}) \le 2$, so G is described in 19.2. In particular $p \le 3$.

Suppose $p = 3$. If G is Co_1 or Co_3 then 23.6 says (2) holds. Otherwise (1) holds by 19.6.

So take $p = 2$. Let $t = z^g$, $P = Q^g\cap H$, and w the width of Q. Then t^* is of type a_2 on $\tilde{Q}$ since $z \in [Q,t] \cong E_8$ by the action of X on V. Further either Q is of width 2 or P^* is extraspecial of width $w-2$, since from the action of X on V, $|P\cap Q| = p^2$ and $P = C_{Q^g}(P\cap Q)$. Assume the latter. If L is HS then as t^* is of type a_2, t^* is not a square in H^*, which is a contradiction as t^* is a square in P^*. Observe also that if $I = Q\cap Q^g$ then $\tilde{I} = [\tilde{Q},t]$, $V = Z(VC_Q(I))$, and Q has two orbits on $V - Q$, so either $C_{H^*}(t^*) = C_H(t)^*$ or $N_H(V)$ is transitive on $V - Q$. In the first case $P^* \trianglelefteq C_{H^*}(t^*)$ and in the second $P^*P^{*s} \trianglelefteq C_{H^*}(t^*)$ where $s \in N_T(V)$ with $[s,t] = z$. We conclude that L is not $M(22)$, and that if G is M_{24}, He, or Co_2 then P^* is weakly closed in T^* while if G is Co_1 or Sz then there is $s \in T$ with $[t,s] = z$ and P^*P^{*s} is weakly closed in P^*. In particular $V = \langle\mathcal{Z}(T)\rangle$ with Q transitive on $\mathcal{Z}(T)$ in the first case and T transitive on $\mathcal{Z}(T)$ in the second. Moreover in the second an argument in the proof of 17.9 shows $O^2(Aut_G(V)) \cong A_6$. Further there is $h \in \langle T,X\rangle$ with $\langle T,Q^h\rangle = \langle T,Q^g\rangle$ and $z^g \in \mathcal{Y}(T)$. Thus (4) holds here. On the other hand in the first case either (3) holds or G is M_{24} or He, and we may assume the latter. Here $N_H(V)^* = C_{H^*}(t^*)$, which is impossible as 3 divides the order of $H\cap X$ but not the order of $C_{H^*}(t^*)$.

As $m(V) = 4$ and $m(G_2(3)) = m(M_{12}) = 3$, G is not $G_2(3)$ or M_{12}. As $3 \in \pi(N_H(V))$ and V^* is of order 2, G is not J_2 or J_3, since there $H^* \cong A_5$ has no element of order 6. Thus we may take Q of width 2. So G is M_{22} or $L_4^{\epsilon}(3)$. In the latter case if $\epsilon = +$ then T has a unique abelian subgroup U of order 16 and $N_G(U)/U \cong A_5$, so $V = U$ and

(3) holds. Similarly if $\epsilon = -$ then there are two choices for U and $N_G(U)/U \cong A_6$, from which we conclude (3) holds. In the former case T has two abelian subgroups V_i of order 16 and $N_G(V_1)/V_1 \cong A_6$ while $N_G(V_2)/V_2$ acts naturally as $PGL_2(4)$ on V_2. Thus (4) holds.

(26.5) Let Γ be the set of subgroups M of G with $Q \le T \cap M \in Syl_p(M)$ and $M \not\le H$. Then one of the following holds:

(1) Q is transitive on $\mathcal{Y}(T)$ and $M = \langle H \cap M, Q^g \rangle$ for each $M \in \Gamma$ and $z^g \in \mathcal{Y}(T)$.

(2) $p = 2$, $G \cong U_4(3)$, Sz, or Co_1, Q is transitive on $\mathcal{Y}(T)$, and $M = \langle H \cap M, Q^g \rangle$ for each $M \le G$ with $T \le M$ and each $z^g \in \mathcal{Y}(T)$.

(3) $p = 2$, $G \cong L_4(3)$ or Co_2, Q is transitive on $\mathcal{Z}(T)$, and $M = \langle H \cap M, Q^g \rangle$ for each $M \in \Gamma$ and $z^g \in \mathcal{Z}(T)$.

(4) $p = 2$, $G \cong M_{24}$ or He, Q has two orbits on $\mathcal{Y}(T)$, and $M = \langle H \cap M, Q^g : z^g \in \mathcal{Y}(T)$ and $Q^g \le M \rangle$ for each $M \in \Gamma$.

(5) $p = 2$ and $G \cong M_{22}$.

(6) $Q \cong 3^{1+4}$.

Proof. We may assume $Q \not\cong 3^{1+4}$ and G is not M_{22}. Let $M \in \Gamma$. By 23.1 and induction on the order of G, the result holds if $O_p(M) = 1$, so we can assume $O_p(M) \ne 1$. Observe that by 19.8 either Q is transitive on $\mathcal{Y}(T)$ or $p = 2$ and G is M_{24} or He. Also by 19.8, $\mathcal{Y}(T)$ is empty in case (3) where Q is transitive on $\mathcal{Z}(T)$ by 26.4. Observe $\mathfrak{D}(M,T\cap Q,Q) \subseteq \mathfrak{D}(G,T,Q)$. With this in mind, by 3.2 we can assume $M \in \mathfrak{D}(G,T,Q)$, unless $p = 2$ and G is in case (2), where at least $M = \langle H \cap M, N : N \in \mathfrak{D}(M,T,Q)$ and $N \not\le H \rangle$. Therefore by 26.3 and 26.4, $M = \langle Q, Q^g \rangle$ where either $z^g \in \mathcal{Y}(T)$ or G is one of the exceptional groups listed in 26.4 and $z^g \in \mathcal{Z}(T)$. Of course in case (2) we can only say this about $N \in \mathfrak{D}(M,T,Q)$, not about M. Thus we have reduced to case (2), where we can assume $N = \langle Q, Q^g \rangle$ for some $z^g \in \mathcal{Z}(T)$. Let $V = \langle z^N \rangle$. From 26.4, $N \le \langle T, Q^h \rangle$ for $z^h \in \mathcal{Y}(T)$, so once again the lemma holds.

27. Geometries.

This section consists of a brief discussion of Tits geometries. The reader is referred to [7] or [38] for more information.

A *geometry* over an index set I is a triple $(\Gamma,\tau,*)$ where Γ is a set of *objects*, $\tau : \Gamma \to I$ is a *type function*, and $*$ is a symmetric *incidence relation* on Γ. The only axiom is: Objects u and v of the same type are incident if and only if $u = v$.

I'll write Γ for the geometry $(\Gamma,\tau,*)$. The *rank* of Γ is the order of I and will always be finite in this paper. Given a subset J of I write J' for the complement I-J to J in I. A *morphism* of geometries is a map of the corresponding object sets preserving type and incidence.

A *flag* of Γ is a set T of objects each pair of which is incident. Because of our one axiom, the type function τ induces an injection of T into I; $\tau(T)$ is called the *type* of T while the orders of $\tau(T)$ and $I-\tau(T)$ are the *rank* and *corank* of T, respectively. Write Γ_T for the set of objects in $\Gamma - T$ incident to each member of T. Also write Γ_T for the geometry

$$\Gamma_T = (\Gamma_T,\tau|_{\Gamma_T},*|_{\Gamma_T})$$

over $I - \tau(T)$. This geometry is called the *residue* of T. Γ_T is said to be a *rank n residue* if T is of corank n. A group G of automorphisms of Γ is *flag transitive* if G is transitive on flags of type J for each subset J of I.

Γ is *connected* if the graph $(\Gamma,*)$ is connected. Γ is *residually connected* if each residue of rank at least 2 is connected and each rank 1 residue is nonempty.

A *generalized n-gon* is a rank 2 geometry of diameter n possessing no cycles of length less than 2n. A *linear space* is a rank 2 geometry on $I = \{1,2\}$ such that for some i in I, each pair of distinct objects of type i is incident with a unique object of type 3-i. The *direct sum* of geometries Γ and Γ' is the geometry with object set $\Gamma \cup \Gamma'$, with induced incidence and type on the subgeometries Γ and Γ', and with each object in Γ incident with each object in Γ'. For example a generalized 2-gon is the direct sum of rank 1 geometries and conversely.

Let G be a group and $\mathcal{F} = (G_i : i \in I)$ a family of subgroups of G. Then we have a geometry $\Gamma(G,\mathcal{F})$ whose set of objects of type i is the coset space G/G_i and with G_ix incident to G_jy if and only if $G_ix \cap G_jy$ is nonempty. Observe G is represented as a group of automorphisms of $\Gamma(G,\mathcal{F})$ via right multiplication.

For $J \subseteq I$ and $x \in$ let $G_J = \bigcap_{j \in J} G_j$, $\mathcal{F}_J = (G_{J \cup \{i\}} : i \in J')$, and $S_{J,x} = (G_jx : j \in J)$. Notice that $S_{J,x}$ is a flag of type J.

(27.1) $\Gamma(G,\mathcal{F})$ is connected if and only if $G = \langle\mathcal{F}\rangle$.

Proof. See 3.2 in [7].

(27.2) The following are equivalent:

(1) G is flag transitive on $\Gamma(G,\mathcal{F})$.

(2) For each $J \subseteq I$ the map

$$\Gamma(G_J,\mathcal{F}_J) \rightarrow \Gamma_{S_J}$$
$$(G_{J\cup\{i\}})x \mapsto G_ix$$

is an isomorphism of the geometry $\Gamma(G_J,\mathcal{F}_J)$ with the residue Γ_{S_J} of the flag S_J.

Proof. See 3.5 in [7].

(27.3) If G is flag transitive on $\Gamma(G,\mathcal{F})$ then $\Gamma(G,\mathcal{F})$ is residually connected if and only if $G_J = \langle\mathcal{F}_J\rangle$ for each $J \subseteq I$ with $|J'| > 1$.

Proof. This is immediate from 27.1 and 27.2.

(27.4) If $\mathcal{F} = (G_1,G_2)$ then $\Gamma(G,\mathcal{F})$ is a generalized 2-gon precisely when $G = G_1G_2$.

Proof. See 3.6 in [7].

The *diagram* of $\Gamma(G,F)$ is the family

$$D(G,\mathcal{F}) = D = (D_J : J \subseteq I,\ |J| = 2)$$

of rank 2 geometries $D_J = \Gamma(G_{J'},\mathcal{F}_{J'})$. The *graph* of the diagram $D(G,\mathcal{F})$ is the graph on I obtained by joining distinct nodes i and j if and only if $D_{i,j}$ is not a generalized 2-gon. By 27.4 this is equivalent to requiring $G_{\{i,j\}'} \neq (G_{i'})(G_{j'})$. D is a *string diagram* if we can order I so that the graph of D is a string:

o——o . . . o——o

(27.5) If $\Gamma(G,\mathcal{F})$ is residually connected and possesses a string diagram then G is flag transitive on $\Gamma(G,\mathcal{F})$.

Proof. See the main theorem of [7].

The following observation is trivial.

(27.6) Assume $\Gamma = \Gamma(G,\mathcal{F})$ is residually connected and G_i is flag transitive on $\Gamma(G_i,\mathcal{F}_i)$ for each $i \in I$. Suppose G is not flag transitive on Γ. Then there exists a flag F of Γ which is not conjugate to a flag S_J for any J. Moreover F is not a flag of $\Gamma(G_i,\mathcal{F}_i)$ for any $G_i \in \mathcal{F}$.

In the tables of the introduction the diagrams for many of the parabolic geometries $\Gamma(G,\mathcal{F})$ are described with pictures such as:

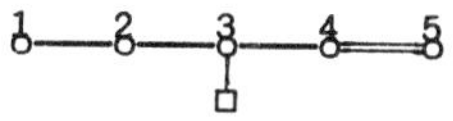

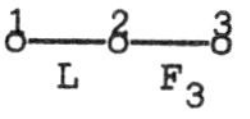

A circle (or *node*) in the picture indexed by the integer i represents the ith type of object. The geometry $D_{i,j}$ is the geometry of the residue of a flag of type $\{i,j\}'$ by 27.2, and the isomorphism type of this geometry defines the nature of the edge connecting the nodes i and j. In particular if there is no edge between i and j then $D_{i,j}$ is a generalized 2-gon. More generally if $D_{i,j}$ is a generalized k-gon then i and j are joined by an edge of weight k-2, written

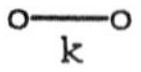

or if $k \leq 2$, writen o——o or o o for $k = 1$ or 2, respectively. Of course this is the usual convention for Dynkin diagrams.

Following Buekenhout in [45], an edge of the form

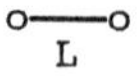

indicates $D_{i,j}$ is a linear space. Given a group H, an edge of the form

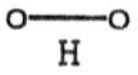

indicates a rank 2 geometry of the form $\Gamma(H,\{H_1,H_2\})$ for suitable subgroups H_1 and H_2, which are hopefully clear from the context.

The squares in the diagram do not represent nodes; rather they represent phantom nodes. That is to say our rank 2 geometry is obtained from a rank 3 or 4 geometry by deleting one or two types of objects. Such a geometry is said to be a *truncation*. Specifically

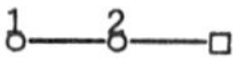

represents the geometry of points and lines of a rank 3 projective geometry,

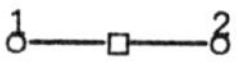

represents the points and hyperplanes of such a geometry,

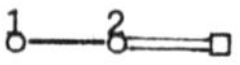

represents the points and lines of the polar space of $Sp_6(2)$ or $P\Omega_7(3)$,

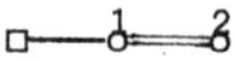

represents the lines and planes of that polar space, and

represents a truncation of the 2-parabolic geometry of M_{24}.

I also use

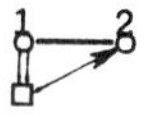

to represent the rank 2 2-parabolic geometry of M_{22}.

The convention of the square node is due to Ronan and Smith [46].

Observe that the nature of the residue Δ of a flag T of type J can be retrieved from the diagram as follows: Delete the nodes in J and all edges through these nodes from the diagram. The resulting diagram E is the sum of connected components E_k, $1 \le k \le r$. The residue Δ is the direct sum of the geometries Δ_k, where Δ_k consists of the objects of Δ whose type is in E_k. Moreover Δ_k is a geometry with diagram E_k.

Also if K is the kernel of the action of G_J on Δ then G_J/K is embedded as a flag transitive subgroup of $Aut(\oplus\Delta_k)$, which is the direct product of the groups $Aut(\Delta_k)$. Often G_J/K is a direct product of obvious subgroups of $Aut(\Delta_k)$. In any case G_J/K is more or less visible from the diagram.

Thus the diagram provides a handy bookkeeping system for recording information about the parabolics and parabolic geometry of G.

28. Parabolics.

Let G be a group, A a group acting on G, and B an A-invariant subgroup of G. An (A,B)-*basis* for G is a family $\mathcal{F} = (G_i : i \in I)$ of overgroups of B in G such that the map $J \mapsto (G_J)^A$ is a bijection of the power set of I with the orbits of A on overgroups of B in G. The basis $\mathcal{F}$ is *regular* if for each $J,K \subseteq I$, $(G_K)^A \cap G_J = (G_K)^{N_A(G_J)}$.

(28.1) Let $\mathcal{F} = (G_i : i \in I)$ be a regular (A,B)-basis for G, let $J,K \subseteq I$, and set $\mathcal{F}_J = \{G_{J \cup \{i\}} : i \in J'\}$. Then

(1) If $(G_K)^a \leq G_J$ for some $a \in A$ then $J \subseteq K$.

(2) $G_J = \langle \mathcal{F}_J \rangle$ if $|J'| > 1$.

(3) $\mathcal{F}_J$ is a regular $(N_A(G_J),B)$-basis for G_J.

(4) $\mathcal{F} \subseteq \mathcal{M}$; that is the members of $\mathcal{F}$ are maximal subgroups of G.

(5) $(G_j : j \in J)$ is a regular $(N_A(G_J),G_J)$-basis for G.

Proof. Let $H = G_J$. Suppose $a \in A$ with $(G_K)^a \leq H$. By regularity of $\mathcal{F}$, $G_K \leq H$. Thus $G_{J \cup K} = G_K \cap H = G_K$, so $K = J \cup K$ as $\mathcal{F}$ is a basis. That is $J \subseteq K$.

The proofs of parts (3) and (5) are straightforward. Then using (3) and proceeding by induction on the order of I, (2) is reduced to (4). But (4) follows from (1) and the basis property.

Let p be a prime and $T \in Syl_p(G)$. Given a subgroup D of Aut(G) containing Inn(G), a D-*twisted p-basis* for G is a family $\mathcal{F} = (G_i : i \in I)$ of representatives for the $N_D(T)$-orbits of maximal overgroups of T in G such that $\mathcal{F}$ is a regular $(N_D(B),B)$-basis for G, where $B = G_I$. The basis $\mathcal{F}$ is *well behaved* if $N_D(T) = N_{D_i}(T)N_{D_j}(T)$ for all distinct i and j in I and D is flag transitive on $\Gamma(D,\mathcal{E})$, where $\mathcal{E} = (D_i : i \in I)$ and $D_i = N_D(G_i)$. A *p-basis* for G is a twisted Inn(G)-basis.

(28.2) Let $\mathcal{F}$ be a D-twisted p-basis for G and $\mathcal{E} = (D_i : i \in I)$. Then

(1) $F = ((D_{i_1})d_{i_1},\dots,(D_{i_k})d_{i_k})$ is a flag in $\mathcal{E}$ if and only if $G(F) = \bigcap_{j=1}^{k} (G_{i_j})^{d_{i_j}}$ contains a Sylow p-subgroup of G.

(2) $N_D(T)$ is transitive on $\mathcal{C}(T) =$

$$\{((G_1)^{d_1},\dots,(G_n)^{d_n}) : d_i \in D \text{ and } T \leq (G_i)^{d_i}\}.$$

(3) Each member of $\mathcal{C}(T)$ is a D-twisted p-basis, so $N_D(T)$ is transitive on D-twisted p-bases over T.

Proof. By flag transitivity of D on $\Gamma(D,\mathcal{E})$, if F is a flag in $\mathcal{E}$ then G(F) contains a Sylow p-group of G. Indeed by flag transitivity, showing (1) amounts to showing that if $d \in D$ and $T \leq G_i \cap (G_j)^d$ then D_i is incident to $(D_j)d$ in $\mathcal{E}$. As T is Sylow in G_j we may take $d \in N_D(T)$. As $\mathcal{F}$ is well behaved, $d = ab$, $a \in N_{D_j}(T)$ and $b \in N_{D_i}(T)$. So $(D_j)d = (D_j)b$ and b maps (D_i,D_j) to $(D_i,(D_j)d)$, establishing the incidence.

Thus (1) holds. Notice (1) and flag transitivity of D on $\Gamma(D,\mathcal{E})$ implies (2) and of course (2) implies (3).

(28.3) Let $\mathcal{F}$ be a well behaved twisted D-basis for G and Y a proper overgroup of T in G. Then

(1) Up to conjugation in $N_D(T)$ there exists a unique parabolic P(Y) of G minimal subject to containing Y.

(2) P(Y) is conugate to $G_{J(Y)}$, where

$$J(Y) = \{i \in I : Y \leq (G_i)^d \text{ for some } d \in D\}.$$

Proof. This is more or less immediate from 28.2.

(28.4) Assume $\mathcal{F} = (G_i : i \in I)$ is a family of proper overgroups of B in G, and $T \leq B \leq H \leq G$ such that:

(a) H is A-invariant with (A,B)-basis $\mathcal{E} = (G_i \cap H : i \in I, G_i \not\leq H)$.

(b) T is A-invariant and there exists a subgroup U of G such that $Y = \langle Y \cap H, U \rangle$ for each overgroup Y of T in G not contained in

(c) There exists $m \in I$ with $H \leq G_m$ and $G_i \not\leq H$ for $i \in m'$.

Then

(1) $H = G_m$ and $\langle B,U \rangle = G_{m'}$.

(2) If $T \leq U$ then U is A-invariant.

(3) $\mathcal{F}$ is an (A,B)-basis for G.

(4) If $\mathcal{E}$ is a regular (A,B)-basis for H then $\mathcal{F}$ is a regular (A,B)-basis for G.

(5) If $T \in Syl_p(G)$, $Inn(G) \le D \le Aut(G)$, and $\mathcal{F}$ is a well behaved D-twisted p-basis for G, then either $P(Y) \le H$ or P(Y) is conjugate under A to $\langle P,U\rangle$ and P is determined up to conjugacy under H subject to being a parabolic of H minimal subject to containing Y.

Proof. Let $\pi : J \mapsto (G_J)^A$ be the map from the power set of I into A-orbits of overgroups of B. By (b), $G = \langle H,U\rangle = M$ for each overgroup M of H in G with $H \ne M$. Thus H is maximal in G and by (c), $H = G_m$.

Also this shows $U \not\le H$. Thus as H is A-invariant, $U^a \not\le H$ for $a \in A$. We may assume $T \le U$. Therefore if $a \notin N_A(U)$ then by (b), $U \le U^a$, contradicting $U \ne U^a$. Thus (2) is established.

For $J \subseteq I$ let $\bar{J} = J - \{m\}$ and $H_J = H \cap G_J$. By (a) and (c), $\mathcal{E} = (H_i : i \in \bar{I})$ is an (A,B)-basis for H. By (b), $G_i = \langle H_i,U\rangle$ for each $i \in \bar{I}$.

Let $B \le Y \le G$. If $Y \le H$ then $Y = (H_{\bar{K}})^a = (G_K)^a$ for some $K \subseteq I$ with $m \in K$ and some $a \in A$. If $Y \not\le H$ then $Y = \langle Y\cap H,U\rangle$ and $Y \cap H = (H_{\bar{K}})^a$, so $Y = \langle H_{\bar{K}},U\rangle^a = (G_K)^a$. Therefore π is a surjection.

As π is surjective, (b) implies $\langle U,B\rangle = G_{m'}$.

Let $J,K \subseteq I$ with $G_J = (G_K)^a$ for some $a \in A$. $G_J \le H$ precisely when $m \in J$, so either

(i) $m \notin J \cup K$ and $G_J \not\le H$, or

(ii) $m \in J \cap K$ and $G_J \le H$.

In either case $(H_{\bar{K}})^a = H_{\bar{J}}$, so by paragraph three of this proof, $\bar{K} = \bar{J}$. But then K = J.

So π is a bijection and (3) is established. Parts (4) and (5) are straightforward.

29. Overgroups of Syow p-groups in groups of GF(p)-type.

In this section assume Q is a large symplectic p-subgroup of the finite group G such that Q is of width at least 2, $L = F^*(G)$ is simple, G = LQ, and L is not of Lie type and characteristic p. Exclude case (2) of Theorem A. Let $Z = \langle z\rangle = \Omega_1(Z(Q))$, $H = N_G(Z)$, and $T \in Syl_p(H)$. Set $H^* = H/Q$ and $\tilde{H} = H/Z$.

In this section we verify Theorem 1 and the tables of the introduction for the groups satisfying the hypothesis just established. For the Mathieu groups this is probably best done using the tables in [19], so we assume L is not a Mathieu group. If $Q \cong 3^{1+4}$ then lemma 26.2 and Theorem 23.1 contain the necessary information, so assume $Q \not\cong 3^{1+4}$. If p = 2 and G is of Lie type over GF(3) then 9.8 is sufficient, so exclude these cases. The case G ≅ He and p = 2 is also a little special:

(29.1) $\mathfrak{A}(T) = \{A_1,A_2\}$ with $A_i \cong E_{64}$ and setting $M_i = N_G(A_i)$, $M_i/A_i \cong S_6/Z_3$. A_1 is fused to A_2 in Aut(G) and $\mathcal{N}(T) = \{H,M_1,M_2\}$.

Proof. Let $H \neq M \in \mathcal{N}(T)$. By 23.1, $O_2(M) \neq 1$. By 26.5, $M = \langle M\cap H, Q^g : Z^g \in \mathcal{Y}(T), g \in M\rangle$ and Q has two orbits on $\mathcal{Y}(T)$. By 20.2, $H \neq M$, so $H\cap M \leq M_i$ for i = 1 or 2, as $\mathcal{N}_H(T) = \{H_i : i = 1,2\}$. Also $Q^g \leq M_1\cap M_2$ for each $Z^g \in \mathcal{Y}(T)$. (cf. 19.9.4) Therefore $M = M_i$.

In the rest of this section we assume G is not He and let $U_0 = \langle Q,Q^g\rangle$, where $Z^g \in \mathcal{Z}(T)$ if G is Co_2 and p = 2, and $Z^g \in \mathcal{Y}(T)$ otherwise. Because of our various exclusions and 26.5 we have:

(29.2) If $M \in \mathcal{N}(T) - \{H\}$ then $M = \langle H\cap M,U_0\rangle$.

We wish to show that if $\mathcal{F} = (G_i : i \in I)$ is a set of representatives for the $N_G(T)$-orbits of maximal overgroups of T, then $\mathcal{F}$ is a well behaved p-basis for G. Lemma 29.2 puts us in a position to apply 28.4. Namely it establishes hypothesis (b) of that result. Hypothesis (c) of 28.4 follows from the maximality of the members of

$\mathcal{F}$. The following result tells us $H \in \mathcal{F}$ and H is $N_G(T)$-invariant:

(29.3) $H \in \mathcal{N}(T)$ and H is $N_{Aut(G)}(T)$-invariant.

Proof. This follows from our exclusions together with 20.2 and 17.1.

Let $\mathcal{E} = (G_i \cap H : i \in I, H \neq G_i)$. To verify hypothesis (a) of 28.4, it suffices by 28.1.5 and 28.3 to know that $\mathcal{E}^* \subseteq \mathcal{N}_{H^*}(T^*)$ and that H^* has a well behaved p-basis. If $H^* \in Chev(p)$ we know the latter fact from the Lie theory. In other cases we obtain this fact inductively or from results in earlier sections. Thus it remains to obtain representatives for the $N_G(T)$-orbits on $\mathcal{N}(T)$ and to verify that if $M \in \mathcal{N}(T) - \{H\}$, then $M \cap H \in \mathcal{N}_H(T)$.

We also need to know that G is flag transitive on $\Gamma(G,\mathcal{F})$. Unless p = 3 and G is ${}^2E_6(2)$ this will follow from 27.5. In the remaining case we will use 32.4.

(29.4) If $M \in \mathcal{N}(T) - \{H\}$ and H is irreducible on $\tilde{Q}$ then $F^*(M)$ is a nonabelian simple group and L and $F^*(M)$ are described in Theorem 23.1.

Proof. This follows from 20.1.

(29.5) Let $U = \langle Z^{U_0} \rangle$. Then

(1) T normalizes U.

(2) $O_p(U_0) = C_{U_0}(U)$, and either $m(U) = 2$ and $Aut_{U_0}(U) = SL(U)$ or G is Co_2, p = 2, $m(U) = 4$, and $Aut_{U_0}(U) \cong \Omega_4^-(2)$.

(3) $U_0 \trianglelefteq N_G(U)$.

(4) $N_G(U) = C_G(U)U_0N_G(T)$.

(5) If $N_H(\tilde{U})$ is maximal in H then $N_G(U)$ is a $N_{Aut(G)}(T)$-invariant maximal subgroup of G.

Proof. By 19.8, 26.4, and our exclusions, Q is transitive on $\mathcal{Y}(T)$ or G is Co_2 and Q is transitive on $\mathcal{Z}(T)$. So (1) and (3) hold. Part (2) follows from 17.6.3 and the definition of $\mathcal{Z}(T)$. Part (4) follows from the earlier parts and a Frattini argument. Part (5) follows from 29.2.

(29.6) Let $U = \langle Z^{U_0}\rangle$ and assume $m(U) = 2$. Then $U = ZZ^g$ for $Z^g \in \mathcal{Y}(T)$, and setting $P = Q^g \cap H$ and letting w be the width of Q, P^* is weakly closed in T^* and normal in $N_{H^*}(\tilde{U})$, and one of the following holds:

(1) $m(Q\cap Q^g) = w+1$ and $P^* \cong E_{p^{w-1}}$.

(2) $p = 2$, G is J_2, J_3, Co_1, or Sz, $Q\cap Q^g = U$, and $P^* = O_2(N_{H^*}(\tilde{U})) \cong E_{2^{2w-2}}$.

(3) $p = 2$, G is HS, $m(Q\cap Q^g) = 3$, and $P^* \cong E_4$.

Proof. Notice that as $U = ZZ^g$, G is not Co_2. If $m(H^*,\tilde{Q}) > 2$ then (1) holds by 19.7. In the remaining cases $p = 2$ and L is HS, M(22), J_2, J_3, Co_1, or Sz by 19.2. If L is M(22) then (1) holds by a remark in the proof of 19.8. If L is HS then $m(H^*) = 2$ and $m(Q) = 3$ or 4 depending on whether G is L or Aut(L). So as $Q\cap Q^g$ is elementary abelian and P is of index 2 in Q^g, it follows that (3) or (1) holds in the respective case. Finally let G be J_2, J_3, Co_1, or Sz and $K = N_H(\tilde{U})$. Then K^* is irreducible on $C_Q(U)/U$ and $O_2(K^*) \cong E_{2^{2w-2}}$. The first remark forces $U = Q\cap Q^g$ and hence also $P^* \cong E_{2^{2w-2}}$. So as $P^* \trianglelefteq K^*$ the second remark shows (2) holds.

(29.7) Assume $m(U) = 2$, let $R = O_p(U_0)$, let w be the width of Q, let $Z^g \in \mathcal{Y}(T)$, and set $V = Q\cap Q^g$. Then $U = Z(R)$ is the natural module for $U_0/R \cong SL_2(p)$, U_0 centralizes V/U, R/V is the direct sum of natural modules for U_0/R, and one of the following holds:

(1) R is of class 3 and order p^{3w-1}. $V = Z_2(R)$. If $N_G(U)$ is irreducible on V/U then R/U is special.

(2) Case (2) of 29.6 holds and R is special of order 4w-2.

(3) $p = 2$ and L is M(22) or HS.

Proof. From 29.5.2, $U = Z(Q\cap R)$ so $U = Z(R)$. The remainder of the initial remarks are just a recapitulation of 17.8. Observe R is of order p^{4w-m}, where $m = m(V)$.

Suppose case (2) of 29.6 holds. Then $U = V$ by 29.6, so as $U = Z(R)$, U_0 is irreducible on U, and R/V is elementary abelian, we conclude (2) holds.

To complete the proof we may assume case (2) of 29.5.5 does not hold and if $p = 2$ then L is not HS or M(22). Then by 19.2 and our exclusions, $m(H^*,\tilde{Q}) > 2$. Now R/U is the product of the abelian groups $(Q\cap H^g)/U$ and $(Q^g\cap H)/U$, so if $Z(R/U) \not\leq V/U$ there is $x \in Q^g\cap H - V$ with $[x,Q\cap H^g] \leq U$. As $x \notin V$, $x^* \neq 1$, while as $|\tilde{U}| = p = |Q:Q\cap H^g|$, we have $m([\tilde{Q},x]) \leq 2$, a contradiction. Therefore $Z(R/U) \leq V/U$. Thus as U_0 centralizes V/U we have $V = Z_2(R)$. It remains to observe that as $V = Z_2(R)$, $U \neq [R,R]$.

From 29.5.5, usually $N_G(U)$ is in our basis $\mathcal{F}$. Indeed often the diagram for $\Gamma(G,\mathcal{F})$ is a string, H is indexed by an end node, and $N_G(U)$ is indexed by the unique adjacent node. Notice also that the structure of $N_G(U)$ can be retrieved almost completely from the information in 29.5, 29.6, and 29.7. In the four cases described in the next lemma these remarks are sufficient.

(29.8) If $p = 2$ and G is J_2 or J_3, $p = 5$ and G is F_5, or $p = 7$ and G is F_1, then $\mathcal{N}(T) = \{H,N_G(U)\}$ and $N_H(U) = N_G(T)$ is maximal in H.

Proof. In each case $N_H(\tilde{U})$ is the unique maximal subgroup of H containing T and not irreducible on $\tilde{Q}$, while by 23.1 each member of $\mathcal{N}(T)$ is local.

Thus we are done with the cases in 29.8 once we see that the structure of $M = N_G(U)$ is as described in the tables of the introduction. But by 29.7, $|U_0|_p = |T|$, so $T \leq U_0$. Then $M = U_0N_G(T)$ by 29.5 and a Frattini argument. A composition series for M is now easily derived from the structure of H and 29.7.

(29.9) If $V = \langle Z^{g_i} : 1 \leq i \leq k\rangle$ with $V \leq \bigcap_{i=1}^{k} Q^{g_i}$ and $M = \langle Q^{g_i} : 1 \leq i \leq k\rangle$, then $\mathrm{Aut}_M(V) = SL(V)$.

Proof Induct on the dimension of V and use 17.8.

It will turn out that G possesses a string diagram unless $p = 3$ and $G \cong {}^2E_6(2)$. We handle that case now along with the case $p = 3$ and $G \cong M(22)$. We saw in section 23 that every member of $\mathcal{N}_H(T)$ is irreducible on $\tilde{Q}$ so if $M \in \mathcal{N}(T)$ with $M\cap H \in \mathcal{N}_H$, then by 29.4, $F^*(M)$ is simple and M is described in 23.1. Further the discussion in section 23 shows there are exactly k such parabolics M_i, where $k = 3$ or 2 for $G \cong {}^2E_6(2)$ or $M(22)$, respectively, and that if Y is an overgroup of T in H which is not irreducible on $\tilde{Q}$ then $Y \le M_i$ for some i. It follows from 29.2 that $\mathcal{N}(T) = \mathcal{F} = \{H, M_i : 1 \le i \le k\}$. So it only remains to establish the flag transitivity of G on $\Gamma(G,\mathcal{F})$ when G is ${}^2E_6(2)$. For this we appeal to 32.4.

Thus in the remaining cases 27.5 supplies the necessary flag transitivity, so we need not consider that question in the remainder of this section. We will also implicitly use an argument in the preceeding paragraph: If $M \in \mathcal{N}(T)$ with $M \cap H$ irreducible on $\tilde{Q}$, then $F^*(M)$ is simple and hence described in Theorem 23.1. So in effect we need only consider members of $\mathcal{N}_H(T)$ which are reducible on $\tilde{Q}$.

If G is F_2 and p = 3 then $H^* \cong O_6^-(2)/Q^3$. Thus there are exactly two maximal subgroups H_i of H containing $O_{3,2}(H)T$ and by 23.1, $H_i = M_i \cap H$ for $M_1 \cong M(23)$ and $M_2 \cong Aut(P\Omega_8^+(3))$. Thus we are done with this case once we observe that $H/Z^*(H)$ does not split over $O_{3,2}(H)/Z^*(H)$, (cf. Theorem 1 in [24]) and hence that each maximal overgroup of T in H contains $O_{3,2}(H)$.

Suppose L is HS and p = 2. Let $I = Z(Q \cap L)$ and $W = \langle I^G \cap T \rangle$. Then $I \cong Z_4$ and hence W is abelian. (cf. 2.5 in [5]) In particular $N_G(W)$ controls fusion in W. $I \le C_Q(U) \le O_2(U_0)$ so $I \ne I^g \le W$, for $Z^g \in \mathcal{Y}(T)$. $I^g \not\le Q$ so $C_Q(U)$ does not centralize I^g so I is not in the center of H. Thus as $K = O^2(H \cap L)$ is of index 2 in $H \cap L$, we have $K = C_L(I)$. In particular K contains W. But then as $N_H(T \cap K)$ is irreducible on $(T \cap K)^*$ and on $C_{Q/I}(T \cap K)$, we have $WQ = T \cap K$ and $(W \cap Q)/I = C_{Q/I}(T \cap K)$. Therefore $W \cong (Z_4)^3$ and $N_L(W)/W \cong L_3(2)$ acts naturally on W with $N_{Aut(L)}(W)$ the split extension of $N_L(W)$ by an involution centralizing each chief section of $N_L(W)$. As $N_G(T \cap K) = \mathcal{N}_H(T)$ it follows that $\mathcal{N}(T) = \{H, N_G(W)\}$, completing the treatment of this case.

Assume G is F_3 and p = 2. Then $H^* \cong A_9$ and $\mathcal{N}_H(T) = \{K\}$ with $K^* \cong A_8$ the stabilizer of a maximal totally singular subspace $\tilde{V}$ of $\tilde{Q}$. Notice $K^* = SL(\tilde{V})$ so each $\tilde{v} \in \tilde{V}^\#$ centralizes some Sylow 2-group of K. Hence as $N_H(V)$ is the unique maximal subgroup of H containing T, $\tilde{V}$ is a TI-set under the action of H. Thus if t^* is an involution in $H^* - K^*$ then $\tilde{V} \cap \tilde{V}^t = 0$, so $m([\tilde{Q},t]) = 4$. Thus $m(H^*,\tilde{Q}) = 4$.

Let $Z^g \in \mathcal{Y}(T)$. Claim $V = Q \cap Q^g$. For $N_H(U)$ acts irreducibly on V/U and $C_Q(U)/V$, so if the claim is false then $C_Q(U)/U = (V/U) \oplus ((Q \cap Q^g)/U)$. But in the latter case $m([\tilde{Q},t]) \le 3$ for $t^* \in O_2(N_H(U))^*$, (as t induces a transvection on V and $\tilde{Q}/V$) contradicting an earlier remark.

As $N_H(U)$ is transitive on V - U, we conclude from the claim and 29.9 that $Aut_G(V) = SL(V)$. Then as $|T| = (|SL(V)|_2)(|V|)$ we conclude $N_G(V)/V \cong L_5(2)$. Of course as $N_H(V) = \mathcal{N}_H(T)$, $\mathcal{N}(T) = \{H, N_G(V)\}$, completing the analysis of this case.

Suppose next that G is Ly and p = 5. Then $H^* \cong Z_2/(Z_4*SL_2(9))$, so there are exactly two maximal subgroups K_i of H containing T and not irreducible on $\tilde{Q}$. $(K_i)^* \cong Z_2/(SL_2(5)*Z_4)$ and by 23.2 there exist $M_i \le G$ with $K_i = M_i \cap H$, $M_1 \cong G_2(5)$, and $M_2/V \cong L_3(5)$, where $V = Q\cap Q^g \cong E_{5^3}$ for $Z^g \in \mathcal{Y}(T)$. Thus $\mathcal{N}(T) = \{H, M_1, M_2\}$, handling this case.

The case $G \cong F_2$ and p = 5 is similar. Here $H^* \cong S_5/(Z_4*QD)$ so again there are just two maximal subgroups K_i of H containing T not irreducible on $\tilde{Q}$. $(K_1)^* \cong Z_4/Z_5/(Z_4*QD)$ and $(K_2)^* \cong Z_2/(Z_4*SL_2(5))$. (As $O^2(H^*) \le Sp_4(5)$ and $A = O^2(H)/Z^*(O^2(H)) \cong A_5/E_{16}$, $O_2(A)$ is the natural module for $A/O_2(A) \cong A_5$ so in particular it has just one class of complements.) By 23.1, $K_1 = M_1 \cap H$ for $M_1 \cong Aut(F_5)$. We use the argument of 23.2 to show $V = Q\cap Q^g \trianglelefteq M_2$ with $M_2/V \cong L_3(5)$ and $Z^g \in \mathcal{Y}(T)$. For this argument to apply we need to know that if t is an involution in U_0 and $W = C_V(t)$ then $W \in Z^G$. But if not then $C_G(W) = W\times B$ with $B \cong Aut(HS)$. This is impossible as $C_{U_0}(W)$ contains a $SL_2(5)/E_{25}$ subgroup, whereas a Sylow 5-group S of HS is 5^{1+2} with Z(S) weakly closed in S by 21.2. This completes the case $G \cong F_5$ and p = 5.

Next let G be F_1 with p = 5. Then $H^* \cong Z_2/((J_2/Z_2)*Z_4)$. So from section 16, there are two classes of maximal overgroups of T in H not irreducible on $\tilde{Q}$ with representatives $K_1 = N_H(T)$ and $K_2 = N_H(X)$, where X is a maximal subgroup of T containing Q with $E((K_2)^*) \cong A_5$. As $N_H(T) \le N_G(U)$ it follows from 29.5 that $N_G(U)$ is maximal in G. $N_H(T)$ is irreducible on $(Q\cap Q^g)/U$ so the structure of $N_G(U)$ is determined by 29.5 and 29.7. Let B be the centralizer in G of a non-2-central involution, Z_0 a 5-central subgroup of B of order 5, and $P = O_5(C_B(Z_0))$. Then $B \cong F_2/Z_2$, $P \cong 5^{1+4}$, and $P \le N_B(P)^\infty$. This forces $Z_0 \in Z^G$, say $Z = Z_0$, and then $P \le Q$. Now by the last paragraph there exists $E_{5^3} \cong V \le Q$ containing Z with $N_B(V)/VZ(B) \cong L_3(5)$. For $b \in N_B(V) - H$, $(Q^b \cap H)Q$ is Sylow in G and of course normalizes V. Thus we may take $V \trianglelefteq T$ and $K_2 = N_H(V)$. Therefore H, $N_G(U)$, and the three $N_G(T)$ conjugates of $M = N_G(V)$ are the maximal subgroups of G containing T. Let $R = O_5(M)$. We've seen $M/R \cong Z_2 \times L_3(5)$. As $|R| = 5^6$, V and R/V are rank 3 irreducibles for M, and from this and the structure of U_0 it is easy to see R is special.

Next let G be M(23) and p = 3. The treatment of this case has some similarity with that of the previous case. Pick t to be a 3-transposition of G centralizing Z and let X of order 3 in Q be inverted by t. Then $N_G(X) = X\langle t\rangle \times A$ with $A \cong \Omega_7(3)$ and $P = Q\cap A = O_3(C_A(Z))$. There is $E_{27} \cong V \le P$ with $N_A(V)$ a maximal parabolic of A and $Aut_A(V) = L_3(3)$. Then $V = \langle Z,Z^g,Z^h\rangle$ for suitable $g,h \in A$, and $Y = \langle Q,Q^g,Q^h\rangle \leq N_G(V)$ by 29.9. $X \leq Q\cap Q^g\cap Q^h = V_1$ and V_1 is abelian so $V_1 \leq C_Q(X) = XP$. Thus $V_1 = X(P\cap P^g\cap P^h) = XV$. Notice Y centralizes V_1/V and $M = N_G(V) = YC_G(V)$ with $C_G(V) \leq N_G(V_1)$, so $V_1 \trianglelefteq M$. There exists an element of Q inducing a transvection on V_1 with axis V, so as Y is irreducible on V, $Aut_M(V_1)$ contains the full group S of transvections with axis V. $|S| = |T|/|N_A(V)X|_3$, so M contains a Sylow 3-subgroup of G which we may take to be T. Further we may take $Z^g \in \mathcal{Y}(T)$.

Let $R = O_3(M)$ and $V_2 = C_R(V_1) = C_R(X) \leq N_A(V)X$. $V_2 \leq X\langle P,P^g,P^h\rangle \leq Y$ and $S = \langle((QV_2/V_2)\cap S)^M\rangle$, so $T = SN_A(V)X \leq Y$ and $M = Y\langle t\rangle$. Further

$$0 < V < V_1 < V_2 \leq R$$

is a Y-chief series for R with V_1/V of rank 1 and all other factors of rank 3. By construction t centralizes all chief factors except $V_1/V \cong X$ and $R/V_2 = S$, and t inverts X. As no element of H^* induces a reflection on $\tilde{Q}$, t inverts S. By 17.7, S is isomorphic to V as a Y/R-module while V_2/V_1 is dual to V.

By 23.1 there exists $N \in \mathcal{N}(T)$ with $N \cong S_3/P\Omega_8^+(3)$. E(N) has 3 parabolics N_i, $1 \leq i \leq 3$, with $O^2(N_i)/R_i \cong L_3(3)$, where $R_i = O_3(N_i) = W_{i1}W_{i2}$, $W_{ij} \cong E_{3^6}$ is the radical of a maximal parabolic N_{ij} over N_i, and $W_{i1}\cap W_{i2} = W_i$ is the natural module for N_i/R_i. Observe $U \leq W_i \leq Q\cap Q^g = W$, and then from+the preceding remarks and the fact that $m(W) = 5$, $W = \langle W_i : 1 \leq i \leq 3\rangle$. Also there is $y \in R - E(N)$ permuting the parabolics N_i transitively. So as W/U is the direct sum of the 3 groups W_i/U, y is free on W/U. This shows that $Z(R/V)\cap(W/V)$ is of order 3. Therefore $Z_2(R)\cap W = V_1$.

If R/V_1 is abelian then so is R^* as $V_1 \leq Q$. But $R^* = O_3((H\cap M)^*) \cong 3^{1+2}$ is nonabelian. (cf. next paragraph) Thus R/V_1 is nonabelian. Therefore as M is irreducible on R/V_2 and V_2/V_1, it follows from the preceeding remark and the preceeding paragraph that R is of class 4, $V_1 = Z_2(R)$, and $V_2 = Z_3(R)$.

It remains to show $\mathcal{N}(T) = \{H,N,M\}$. However $H^* \cong GL_2(3)/3^{1+2}/Q^3$ with $O_{2,3}(H^*)$ irreducible on $\tilde{Q}$ and $E = T\cap O_{3,2,3}(H)$ irreducible on $O_2(H^*)$. Thus $N_H(E)$ is the unique maximal subgroup of H containing T and not irreducible on $\tilde{Q}$. Therefore as $|N_H(E)| = |H\cap M|$ it follows that $N_H(E) = H\cap M$ and M is the unique local maximal subgroup of G containing T, completing the treatment of this case.

(29.10) Let $G = M(24)'$ and $p = 3$. Let $A = Aut(G) = M(24)$ and D the set of 3-transpositions of A. Let $X = x_1 \times {}_2$ where x_i is inverted by $d_i \in D$ and centralized by d_{3-i}. Let $V = O_3(N_G(\langle x\rangle))$. Then $V \cong E_{3^7}$ and $N_G(V)/V \cong \Omega_7(3)$ acts naturally on V with $z^G \cap V$ the set of singular points of the orthogonal space V.

Proof. By [23], $C_A(x) = X \times B$ where $X = \langle x\rangle$, $B = \langle C_D(x)\rangle$, $O_3(B) = P \cong E_{3^6}$, and $B/P \cong Z_2/\Omega_6^-(3)$ acts naturally on P. Let Z_0 be a singular point of P and T_0 a Sylow 3-subgroup of $N_B(Z_0)X$. Without loss $T_0 \le T$. Then as $XZ_0 = Z(T_0)$ and $Z_0 = XZ_0 \cap [T_0,T_0]$, we have $Z_0 = Z(T) \cap T_0 = Z$. Further $V = J(O_3(N_B(Z))X)$ and $|O_3(N_B(Z))X| = |Q|$, so $Q \le N_G(V)$ and $|V:V\cap Q| = 3$. Finally $\langle B,Q\rangle$ is irreducible on V so 17.6.4 completes the proof of the lemma.

Lemma 29.10 allows us to handle the case $G = M(24)'$ and $p = 3$. Let V be the subgroup of G described in 29.10. Then $I = N_G(V)$ contains a Sylow 3-group of G which we may take to be T. Observe T is contained in $|N_G(T):N_I(T)| = 2$ conjugates of $N_G(T)$.

Next $H^* \cong Z_2/U_5(2)$. So by 2.4, up to conjugacy in H there are exactly 3 maximal subgroups H_i of H containing T and not irreducible on $\tilde{Q}$. $(H_1)^* \cong Z_2/(Z_3 \times U_4(2))$, $(H_2)^* \cong Z_2/(U_2(2) \times GU_3(2))$, and $(H_3)^* \cong (Z_2 \times S_5)/E_{81}$. Then $H_1 = N_H(V)$ and $H_3 = N_H(U)$. So $N_G(V)$ and $N_G(U)$ are maximal in G. The structure of $N_G(U)$ described in the tables of the introduction can be retrieved from 29.7 and the observation that $O_3(U_0) = O_3(N_G(U))$. This last remark follows from the fact that $|O_3(H_3)| = 3^{15} = 3|O_3(U_0)|$ by 29.7.

Finally let d be a 3-transposition in $A = \mathrm{Aut}(G)$ and $K = C_G(d)$. Then $K \cong M(23)$, so by our treatment of the case (M(23),3) there is a subgroup M_K of K such that $M_K/S \cong GL_3(3)$ and $S = O_3(M_K)$ is of order 3^{10}. Indeed setting $E = Z_2(S)$ and $W = Z(S)$, there is X of order 3 in $E - W$ inverted by a 3-transposition in $K\langle d\rangle$ such that $M_K \cap C(X) \cong L_3(3)/E_{27}/E_{27}$. As X is inverted by a 3-transposition, $N_A(X) \cong S_3 \times (S_3/P\Omega_8^+(3))$ and d induces a reflection on $B = E(C_G(X))$. In particular $(M_K)^\infty$ is a parabolic in $E(C_B(d)) \cong \Omega_7(3)$ and $M_X = N_B(W)^\infty$ is a parabolic in B with $R_X = O_3(M_X)$ special of order 3^6. Let $M = N_G(E) \cap N_G(W)$. Then $\langle M_K,R_X\rangle \le M$ with $R_3 = XR_X = C_G(E)$ and S inducing the group of transvections on E with axis W, so $R = SR_X \trianglelefteq M$. $|M|_3 = 3^{16}$ so M contains a Sylow 3-group of G which we may take to be T. As $|H\cap M| \ge |H_2|$, $H_2 = H \cap M$ and $M \in \mathcal{N}(T)$. Notice $R = O_3(M)$. The 3-chief factors of M are visible in R_X and S; there are 4 of rank 3 and 1 of rank 1. Also $R_3 = VV^y$ for $y \in M_K - I$. So as R centralizes V/E, $R_3 \le Z_3(R)$. Therefore as S is of class 4 with $W = Z(S)$, $E = Z_2(S)$, and $XC_S(X) = Z_3(S)$, it follows that R is of class 4 with $W = Z(R)$, $E = Z_2(R)$, and $R_3 = X \times R_X = Z_3(R)$.

Assume next that G is F_1 and $p = 3$. Then $H^* \cong Z_2/Sz/Z_2$. By 11.7, up to conjugacy in H there are three members H_i of $\mathcal{N}_H(T)$ not irreducible on $\tilde{Q}$. Further $H_1^* \cong E_4/U_4(3)/Z_6$, $H_2^* = N_{H^*}(Z(T^*)) \cong Z_2/(SL_2(3)*D_8)/3^6/Z_2$, and $H_3^* = N_{H^*}(J(T^*)) \cong (Z_2 \times M_{11})/E_{3^5}/Z_2$.

Let X be of order 3 in G with $K/X \cong M(24)$ and $N_T(X) \in Syl_3(K)$, where $K = N_G(X)$. We conclude from the structure of $N_K(Z)$ that $X \le Q$ with $C_Q(X) = O_3(N_K(Z))$. By the previous case there is $M_X \le K$ with $O_3(M_X) = V \cong E_{3^8}$ and $M_X/V \cong Z_2/\Omega_7(3)$. As in an earlier case, $V = J(C_Q(X)V)$ and $|Q| = 3|C_Q(X)V|$, so $Q \le M = N_G(V)$ and $|V:Q\cap V| = 3$. Then by 17.6.4, V is an orthogonal space for $O^2(M/V) \cong P\Omega_8^{\epsilon}(3)$. Then $T \le M$ by an order argument, and as $O^2(H\cap M)^* \cong \Omega_6^{\epsilon}(3)/Z_3$, it follows that $O^2(H\cap M) = O^2(H_1)$. In particular $\epsilon = -$ as $O^2(H_1^*) \cong \Omega_6^-(3)/Z_3$. Now $V = J(O_3(H_1)) \trianglelefteq H_1$ so $H_1 = H\cap M$, $M \in \mathcal{N}(T)$, and $M/V \cong Z_2/\Omega_8^-(3)$.

Next $(Q^g\cap H)^* \cong E_{3^5}$ by 29.6, so $J(T^*) = (Q^g\cap H)^*$. Hence $H_3 = N_H(U)$ and $N_G(U) \in \mathcal{N}(T)$. As usual the structure of $N_G(U)$ can be retrieved from 29.5 and 29.7.

Let W be a T-invariant totally singular 3-subspace of V, $I = N_G(W)$, and $R = O_3(I)$. Then $I\cap H\cap M = H_1\cap H_2$ is of index 2 in H_2 and there exists $y \in N_{K\cap H}(W) - M_X$, so $H_2 = I\cap H$ and $I \in \mathcal{N}(T)$. In particular $\mathcal{N}(T) = \{H, N_G(U), I, M\}$. Observe also that $|I:I\cap M| = 2$ and $R = O_3(I\cap M)$.

Let $I_X = K\cap I$ and $S = O_3(I_X)$, $E/X = Z_2(S/X)$, and $S_3/X = Z_3(S/X)$. Then from our discussion of M(23) and M(24), $E = Q\cap Q^g\cap Q^h$, where $W = \langle Z, Z^g, Z^h\rangle$, and $|E:W| = 9$. Let $J = I^{\infty}$. Notice $I = JC_I(W)$ so $I/R = (J/R) \times C_I(J/R)/R$. From the structure of H_2, $C_I(J/R) = RP$ with $P \cong D_8/Z_2$. R is transitive on subgroups of order 3 in XW - W, so $C_I(E/W) = RC_{I\cap M_X}(E/W) = J$. Therefore P is faithful on E/W, so P is a Sylow 2-group SD_{16} of $GL_2(3)$. That is $I/R \cong L_3(3) \times SD_{16}$.

Also $C_R(E) = S_3$ is of index 3^6 in R with S/S_3 J/R-isomorphic to W, so as P is irreducible on E/W it follows that R/S_3 is the $SD_{16}\otimes L_3(3)$ module for I/R. Next $I = RN_I(Y)$ for some P-invariant complement Y to W in E containing X. Moreover $N_K(Y)/Y \cong Z_2 \times (S_3/P\Omega_8^+(3))$ and P is faithful on Y and V/Y, so $E(C_G(Y)) = D \cong P\Omega_8^+(3)$, $C_G(D)\cap N(Y) = YZ(P)$, and $N_G(Y)/YZ(P) \cong Aut(P\Omega_8^+(3))$. In particular visible inside $N_G(Y)$ is the fact that S_3/E is the $D_8 \otimes L_3(3)^{dual}$ module for I/R. We conclude from this and previous remarks that R is of class 4 with $W = Z(R)$, $E = Z_2(R)$, and $S_3 = Z_3(R)$.

This completes the treatment of all cases with p odd. It remains only to consider certain case where $p = 2$ and $\Gamma(G,\mathcal{F})$ is of high rank. These cases are dealt with in the next section.

30. Overgroups of large symplectic 2-subgroups.

In this section we continue the hypothesis and notation of the last section with p = 2. In particular $U = \langle Z^{U_0}\rangle$ and its normalizer are described in 29.5, 29.6, and 29.7. We have reduced to the case where G is Sz, Co_2, Co_1, J_4, F_5, Aut(M(22)), M(24)′, F_2, or F_1.

(30.1) Let G be J_4 or M(24)′. Then

(1) There exists X of order 3 in H with $XQ \trianglelefteq H$. Thus $\tilde{Q}$ is an irreducible $GF(4)C_H(X)$-module.

(2) Let $z^g \in \mathcal{Y}(T)$ and $\tilde{W}$ the 1-dimensional GF(4)-subspace of $\tilde{Q}$ containing z^g. Then $W \leq Q \cap Q^g$ and $Aut_G(W) = L_3(2)$.

Proof. The first remark is evident from the structure of H. For the second observe that $Q^g \cap H$ centralizes W/U so as $m(H^*, \tilde{Q}) > 2$, $W \leq Q^g$. Then as XQ is transitive on W - Z, the proof is complete.

The observations in the next lemma are contained in Janko's treatment of J_4. A proof of these observations is included for completeness.

(30.2) Let G be J_4. Then

(1) H is transitive on elements of order 4 in Q and if y is such an element then $C_G(y)^* \cong PGL_2(11)$.

(2) Let A be of order 11 in $C_G(y)$ and X of order 3 in $C_H(A)$. Then $S = C_Q(A) \cong Q_8$, $XQ \trianglelefteq H$, $Z = C_Q(X)$, and $C_G(S)^{\infty} \cong L_2(11)/Q^5$.

(3) $V = J(T) \cong E_{2^{11}}$ and $N_H(V)^* \cong S_6/E_{16}$.

(4) Let $w = z^g \in \mathcal{Y}(T)$. Then $C_H(w)^* = N_H(J(T^*))^* \cong S_5/E_{32}$.

(5) $N_G(V)/V \cong M_{24}$ with V the Todd module for $N_G(V)/V$.

Proof. $Q \cong D^6$ so $H^* \leq O^+_{12}(2)$ and there are $2^5(2^6-1)$ nonsingular points in $\tilde{Q}$. From the action of $O^+_{12}(2)$ on $\tilde{Q}$, $C_{\tilde{Q}}(A)$ is nondegenerate of sign -1, so $S = C_Q(A) \cong Q_8$. Let $y \in S - Z$. Thus y is of order 4. As $H^* \cong Aut(M_{22})/Z_3$, a subgroup X of order 3 in H centralized by A is Sylow in $O_{2,3}(H)$ and setting $K = C_H(X)$, K/Z is either quasisimple and isomorphisc to M_{22}/Z_3 or isomorphic to $Z_2/M_{22}/Z_3$. As a matter of fact it is the former, although we don't need that fact. In any event K is irreducible on $\tilde{Q}$ so $C_Q(X) = Z$. From [19] there is a unique maximal subgroup K_0 of K containing AZ; further $K_0 = XZxK_1$ where $K_1 = (K_0)^\infty \cong L_2(11)$. Then $C_K(\tilde{y}) \le K_1Z$ so as $|K:K_1Z| = 2^5(2^6-1) \ge |\tilde{y}^K|$, it follows that $ZK_1 = C_K(y)$ and H is transitive on the elements of Q of order 4. Further K_1 centralizes $[X,y] = S$, so (2) holds.

Let $E_{32} \cong B \le K$ with $N_K(B)/B \cong A_6/Z_3$. Let $P \in Syl_3(N_K(B))$. Then $P \cong 3^{1+2}$ with $X = Z(P)$. So as $C_{\tilde{Q}}(X) = 0$, each irreducible for P on $\tilde{Q}$ is of rank 6. It follows that $m(C_{\tilde{Q}}(B)) = 6$ and $C_{\tilde{Q}}(B) = C_Q(B)/Z$. Moreover as P is irreducible on $C_{\tilde{Q}}(B)$, $C_{\tilde{Q}}(B)$ is totally singular or nondegenerate. In the latter case B also centralizes $C_{\tilde{Q}}(B)^\perp$, a contradiction. Therefore $C_Q(B) \cong E_{2^7}$.

Observe also that for $b \in B - Z$, b is not fused to bz and b is of type a_4 on $\tilde{Q}$ from the action of K on $\tilde{Q}$. Thus setting $H_M = N_H(B^*)$, $|H_M:H_M \cap C_H(b)| = 240$. Further if $v \in C_Q(B) - [Q,b]$ then $C_{H_M}(bv) = H_M \cap C(b) \cap C(v)$ is of index 6 in $C_{H_M}(b)$.

Let $V = C_Q(B)B$. We've shown $V \cong E_{2^{11}}$. Suppose J is an elementary abelian subgroup of T of rank at least 11. Then $m(J \cap Q) \le m(Q) = 7$ and $m(J^*) \le m(H^*) = 5$. Moreover if $m(J^*) = 5$ then $J^* = J(T^*) \not\le C(X^*)$. But for $j \in J^* - C(X^*)$, j inverts X^* and hence $m(C_{\tilde{Q}}(j)) = 6$ and $C_{\tilde{Q}}(j) \not\le C(i)$ for $i^* \in J^* - \langle j^* \rangle$. We conclude $J^* \le C(X^*)$, $m(J^*) = 4$, $m(J \cap Q) = 7$, and $m(J) = 11$. Now $\mathcal{A}(C_T*(X^*)) = \{B^*, C_{J(T^*)}(X^*)\}$. Also by 29.7, $E^* = (Q^g \cap H)^* = J(T^*)$, where $z^g \in \mathcal{Y}(T)$. By 19.7, $C_{\tilde{Q}}(E) = \tilde{U}$, so $m(C_{\tilde{Q}}(C_E*(X^*))) \le 2$. Therefore $J^* = B^*$. Then $J \cap Q = C_Q(B) = V \cap Q$ and hence $J = C_G(V \cap Q)) = V$. Moreover $V = J(VQ) \trianglelefteq N_H(V^*)$ so $H_M = N_H(V)$. Therefore (3) and (4) hold.

Let's calculate the orbits of H_M on V. First z is an orbit of length 1. As $(Q \cap V)/Z$ is the 3-dimensional GF(4)-module for $(H_M)^\infty/O_2(H_M) \cong A_6/Z_3$, H_M has two orbits on $Q \cap V - Z$ of length 90 and 36, with z^g in the first orbit. Next H_M is transitive on $(V/(Q \cap V))^\#$, so each member of $V - Q$ is fused into bV. Also X is transitive on $((Q \cap V)/[Q,b])^\#$, so the orbits of H_M on $V - Q$ have representatives b, bz, and bv, and by an earlier remark are of length 240, 240, and 1440.

Claim 7 divides the order of $M = N_G(V)$. For define W as in 30.1. Then $W \le V$ so by a Frattini argument, $N_G(W) = C_G(W)N_M(W)$. Now 30.1.2 completes the proof of the claim. Now as 7 does not divide the order of H_M, $|z^M| \equiv 0 \bmod 7$. From this it follows that z^M consists of the orbits of length 1,90, and 1440 plus one of the 240 orbits, and that $|z^M| = 7.11.23$. Therefore M/V has the same order as M_{24}. From this and the fact that $H_M/V \cong S_6/Z_3/E_{64}$ we conclude that M is irreducible on V and M/V is simple, and then conclude $M/V \cong M_{24}$. Finally by 31.4, V is the Todd module.

(30.3) Let G be J_4, M(24)′, Co_1, Co_2, or Aut(M(22)), and let V = J(T). Then

(1) $V \cong E_{2^{11}}$ and $N_G(V)/V \cong M_{24}$ if G is J_4, M(24)′, or Co_1, and $V \cong E_{2^{10}}$ and $N_G(V)/V \cong \mathrm{Aut}(M_{22})$ if G is Co_2 or Aut(M(22)).

(2) V is the Todd module for $N_G(V)/V$ if G is J_4, M(24)′, or Aut(M(22)) while V is the dual Todd module if G is a Conway group.

Proof. We've already established this fact if G is J_4. If G is a Conway group or a Fischer group, then from the representation of G on the Leech lattice or on its set of 3-transpositions, there is an elementary 2-group V such that $N_G(V)$ is as claimed. It remains to observe that $V = J(N_G(V))$ from 31.9, and hence V = J(T). So 30.3 holds.

Assume that G is J_4 or M(24)′. We determine $\mathcal{N}(T)$ in each case. Let $z^g \in \mathcal{Y}(T)$ and $E = Q^g \cap H$. Define X and W as in 30.1, let V = J(T), and $M = N_G(V)$.

Arguing as in 30.2, $E^* = J(T^*)$. Notice $N_H(W) \in \mathcal{N}_H(T)$ while by 30.1, $\mathrm{Aut}_G(W) = L_3(2)$. In particular $N_G(W) \in \mathcal{N}(T)$ by 29.2. $[E, Q \cap Q^g] = z^g$, so as X does not act on U, X does not act on $Q \cap Q^g$. On the other hand an element of order 5 in $N_H(W)$ has two rank 4 chief factors on $C_Q(W)/W$, and so is irreducible on $(Q \cap Q^g)/W$. It follows that $Q \cap Q^g \cap Q^{gx} = W$ for $x \in X^\#$. Then from 17.7 and the structure of H, $N_G(W)$ is as claimed in the introduction.

If G is J_4 then from the list of maximal subgroups of H^* in [19] and 30.2, $\{N_H(W), N_H(V)\} = \mathcal{N}_H(T)$. Thus $\mathcal{N}(T) = \{H, N_G(W), N_G(V)\}$ and we are done. So take G to be M(24)′. Then $H^* \cong Z_2/U_4(3)/Z_3$, so $\mathcal{N}_H(T)$ is listed in the introduction by 9.8. In particular in addition to $N_H(W)$ and $N_H(V)$ there is just one more member of $\mathcal{N}_H(T)$, and that is $H_3 = C_H(Z(T^*))$. By 30.3 and 31.7 there is a T-invariant rank 6 subspace D of V with $N_M(D)/V \cong A_8/E_{16}$ and with D the $\Omega_6^+(2)$-module for $N_M(D)/O_2(N_M(D))$. Then $Q \cap D$

= [Q,D] is of index 2 in D so $Z_2 \cong D^* \trianglelefteq T^*$ so $D^* = Z(T^*)$ and $D\cap Q = [Q,D]$. Thus H_3 acts on DQ and $D\cap Q$, and then also on $Z(C_{DQ}(D\cap Q)) = D$. Therefore $\mathcal{N}(T) = \{H, N_G(W), N_G(D), N_G(V)\}$. Moreover $N_M(D)$ is irreducible on the 4-dimensional spaces $O_2(N_M(D))/V$ and $[V, N_M(D)]/D$ while $C_G(D)/O_2(C_G(D)) \cong S_3$ and is nontrivial on $O_2(N_G(D))/D$. From these remarks we conclude $N_G(D)$ is as claimed in the introduction, completing the treatment of this case.

The case $G \cong F_5$ is similar. Here $H^* \cong Z_2 wr A_5$ and $\tilde{Q}$ is the $\Omega_4^+(4)$-module for $E(H^*)$. Let $\tilde{W}$ be the 1-dimensional GF(4)-space with $z^g \in W$ and let $E = Q^g \cap H$. The argument of 30.1 shows $Aut_G(W) \cong L_3(2)$ and $W \leq Q\cap E$. In particular $Aut_E(W) \cong Z_2$ is nontrivial on $\tilde{W}$ so there is $e^* \in E^* - E(H^*)$ and $E^* = C_{T^*}(e)$. Also $N_H(W) = N_H(S)$, where $S^* = T^* \cap E(H^*)$, and $\mathcal{N}_H(T) = \{N_H(S)\}$. Therefore $\mathcal{N}(T) = \{H, N_G(W)\}$. $N_H(W)/O_2(N_H(W)) \cong Z_3 \times S_3$ so $N_G(W)/R = (C_G(W)/R) \times (N_G(W)^\infty/R) \cong Z_3 \times L_3(2)$, for $R = O_2(N_G(W))$.

Let $\tilde{W}^\perp$ be the subspace orthogonal to $\tilde{W}$ in $\tilde{Q}$ regarded as a 4-dimensional orthogonal space over GF(4). Then an element of order 3 in $C_H(W)$ is fixed point free on $W^\perp/W$, so $W^\perp \leq C_Q(W)$. Also $N_H(S) = C_H(W^\perp/W)C_H(W)\langle e\rangle$. Thus $C_H(W^\perp/W)$ is trnasitive on $\tilde{W}^\#$ and acts on $W[W^\perp,e]$. But $e \in E$ and $W^\perp \leq C_Q(W)$, so $[W^\perp,e] \leq Q\cap Q^g$. Then as $C_H(W^\perp/W)$ is trnasitive on $W^\#$, $W[W^\perp,e] \leq Q\cap Q^g \cap Q^h$ for $W = \langle z, z^g, z^h\rangle$. Therefore as $\tilde{W}[\tilde{W}^\perp,e]$ is of rank 4 and $Q\cap Q^g$ is abelian, and hence of rank at most 5, we conclude $Q\cap Q^g = Q\cap Q^g\cap Q^h = W[W^\perp,e]$. The structure of $N_G(W)$ can be retrieved from these remarks, the preceeding paragraph, and 17.7.

There are also similarities in the case $G = Aut(M(22))$. Here $H^* \cong O_6^-(2)$ and $\tilde{Q}$ has an H^*-composition series 1/8/1 with the noncentral chief factor the 4-dimensional $U_4(2)$-module over GF(4). Let $V = J(T)$ and $M = N_G(V)$. Then $M/V \cong Aut(M_{22})$ with V the Todd module by 30.3. Hence by 31.8, $(H\cap M)/O_2(H\cap M) \cong S_5$, so $S_5/E_{16} \cong (H\cap M)^*$ and $\mathcal{N}_H(T) = \{H\cap M, H_1\}$, where $H_1 = C_H(Z(T^*))$.

By 31.8, there is a T-invariant 4-dimensional subspace D of V with $N_M(D)/V \cong S_6/E_{16}$ and D the natural module. By 17.6.6, $D \not\leq Q$, so as $Q\cap D \trianglelefteq N_{H\cap M}(D)$, $Q\cap D$ is a hyperplane of D. By an argument in the treatment of M(24)′, $D^* = Z(T^*)$, $[D,Q] = D\cap Q$, and H_1 acts on $D = Z(C_{DQ}([D,Q]))$. So $\mathcal{N}(T) = \{H, N_G(D), M\}$.

Let $K = N_G(D)$ and $R = O_2(K)$. Recall from the discusion in 19.8 that z is the product of commuting 3-transpositions a and b with $\langle a,b\rangle \trianglelefteq H$. Then $\langle a,D\rangle = \langle a^K\rangle \trianglelefteq K$. Let $I = \langle Q^K\rangle$. Observe $I = \langle Q^{K\cap M}\rangle$ and $K/R = (I/R) \times (C_G(D)/R) \cong S_6 \times S_3$. The structure of R described in the introduction can now be retrieved from $K\cap H$ and $K\cap M$.

Let G be Co_2. Remember this is the only case where $m(U) = 4$ and $z^g \in \mathcal{Z}(T)$. Let $M = N_G(U)$ and $R = O_2(M)$. Let $V = J(T)$. By 30.3 and 31.8, $N_H(V)/O_2(N_H(V)) \cong S_6$. But $H^* \cong Sp_6(2)$ with $\tilde{Q}$ the spin module for H^*, so $\mathcal{N}_H(T) = \{H_1,H_2,H_3\}$ where $H_1^* \cong S_6/E_{32}$ stabilizes a maximal totally singular subspace $\tilde{E}$ of $\tilde{Q}$, $H_2^* \cong S_3 \times S_3/2^7$ is the stabilizer of a singular line of $\tilde{Q}$, and $H_3^* \cong L_3(2)/2^6$ is the stabilizer of a singular point of $\tilde{Q}$. We conclude $H_1 = N_H(V)$ and $N_G(V) \in \mathcal{N}(T)$. Arguing as we have already several times, $Z_2 \cong U^*$ is generated by the unique element of type a_2 in $Z(T^*)$ and $H_2 = N_H(U)$ with $U = Z(C_{UQ}(C_Q([U,Q])))$. So $M \in \mathcal{N}(T)$. $M/R = (Z_2/U_0/R) \times (C_G(U)/R) \cong S_3 \times S_5$ and the action of M/R on R is visible in $N_G(V)$ and H. To complete this case it remains only to observe that $G = \langle H_3,U_0\rangle$. This is because $D^* = \langle U^{H_3}\rangle^* = O_2(H_3^*) \cong E_{64}$ and $\langle (U\cap Q)^{H_3}\rangle = E \cong E_{32}$, so if D is abelian then $m(D) \geq 11$. But by 30.3, $m(G) = 10$. Also $D \leq \langle z^{\langle H_3,U_0\rangle}$ so if D is not abelian then by 17.4, $F^*(\langle H_3,U_0\rangle)$ is simple, so $G = \langle H_3,U_0\rangle$ by 23.1.

We have reduced to the case $G \cong Sz$, Co_1,F_2, or F_1. In each of these cases H is indexed by an end node of the diagram of G (Say by node 1.) and $N_G(U) \in \mathcal{N}(T)$ is indexed by the adjacent node (Which must then be node 2.) To see that $N_G(U) \in \mathcal{N}(T)$ we need $N_H(T) \in \mathcal{N}_H(T)$. If G is Sz or Co_1 this is clear as H^* is the commutator group of the orthogonal group on $\tilde{Q}$ so the stabilizer of the singular point $\tilde{U}$ is a maximal parabolic. If G is F_1 or F_2 then $E^* = (Q^g \cap H)^* \cong E_{2^{w-1}}$ by 29.6. But $H^* \cong Co_1$ or Co_2 and $w-1 = 11$ or $10 = m(H^*)$ by 30.3. So again by 30.3 and the treatment of the Conway groups, (in an inductive context) $E^* = J(T^*)$ and $N_H(U)^* \cong M_{24}/E^*$ or $Aut(M_{22})/E^* \in \mathcal{N}_H*(T^*)$.

Observe that the structure of $N_G(U)$ is supplied by 29.5 and 29.7. Let $R = O_2(N_G(U))$. If G is F_1 or F_2 then $R^* = E^*$ is the dual Todd module while in the other two cases it is the natural module for the appropriate orthogonal group. Then as $m(R/\Phi(R)) = 2m(E^*)$ by 29.6 and 29.7, $R/\Phi(R)$ is the tensor product of the 2-dimensional module for $L_2(2) \cong U_0/O_2(U_0)$ with the module E^* for $N_H(U)/O_2(N_H(U))$. Also $(Q\cap Q^g)/U$ is dual to $C_Q(U)/(Q\cap Q^g)$, which is in turn $N_H(U)$-isomorphic to E^*. Thus the structure of $N_G(U)$ is as claimed in the introduction.

Let G be Sz and $\tilde{D}$ a singular line of $\tilde{Q}$. Then $\mathcal{N}_H(T) = \{N_H(D),N_H(U)\}$. Set $V = \langle D^{U_0}\rangle$, $I = N_G(U)$, and $S = O_2(I)$. Then from the structure of I, $V \cong E_{16}$, $N_I(V)/S \cong S_3 \times A_4$, and $|S:C_S(V)| \le 16$. As there are no $N_I(V)$-invariant subgroups of S/U of index 8 or 16 it follows that $|S:C_S(V)| = 4$. Therefore $V = C_G(C_Q(D)) \trianglelefteq N_H(D)$ so $\mathcal{N}(T) = \{H, I, N_G(V)\}$. The structure of $N_G(V)$ is easily determined from that of $N_H(D)$ and I, and from 17.6.

Let G be Co_1, $V = J(T)$, and $M = N_G(V)$. From 30.3, V is the dual Todd module for $M/V \cong M_{24}$. So by 31.7, $(H\cap M)/O_2(H\cap M) \cong A_8$. As $H^* \cong \Omega_8^+(2)$ this forces $(H\cap M)^*$ to be a maximal parabolic of H^* and $M \in \mathcal{N}(T)$. Next we repeat an argument just used for Sz to find $E_{16} \cong W$ with $\tilde{D}$ the T-invariant singular line of $\tilde{Q}$, $D = C_G(C_Q(D)) \trianglelefteq N_H(D)$, and $U_0 \le I = N_G(V)$. Then $I \in \mathcal{N}(T)$ and the structure of I is easily recovered. To complete this case it remains to show $G = \langle U_0, K\rangle$, where K^* is the remaining maximal parabolic of H^*. Let $X = \langle U_0, K\rangle$ and suppose X is proper in G. Then by 23.1, $E = \langle z^M\rangle$ is abelian. As $Q\cap Q^g = U$ by 29.3, $E \not\le Q$. But now an argument used in the Co_2 case applies.

This leaves $G \cong F_2$ or F_1. The geometry for G is of rank n = 4 or 5. We first construct the parabolic M of G indexed by the nth node of the diagram. Let e be a non-2-central involution of G with $C_T(e)$ Sylow in $G_e = C_G(e)$. For $X \le G$ write X_e for $X\cap G_e$. Then G_e is isomorphic to $Z_2/{}^2E_6(2)/Z_2$ or F_2/Z_2, respectively. In the latter case the parabolics of G_e are known in an inductive context. Also $H_e \cong Z_2/U_6(2)/(Z_2 \times D^{10})$ or $Co_2/(Z_2 \times D^{11})$ and G_e has a maximal parabolic M_e containing T_e and isomorphic to $O_8^-(2)/E_{2^{16}}/(Z_2 \times E_{2^8})$ or $Sp_8(2)/E_{2^{16}}/(E_4 \times E_{2^8})$, respectively. Let $V_0 = \langle z^{M_e}\rangle$. Let $V = V_0$ if G is F_2 and let $V = Z(O_2(M_e)) \cong E_{2^{10}}$ if G is F_1.

Observe that $m(V_0) = 8$ if G is F_2 and $m(V_0) = 8$ or 9 if G is F_1. In the first case this holds as the 1-cohomology of $\Omega_8^-(2)$ on its natural module is trivial. Also observe that $Q\cap V_0$ is a hyperplane of V_0 and $V = V_0(Q\cap V)$. Finally notice that VQ_e/Q_e is a 2-central involution in H_e/Q_e, so V^* is a 2-central involution of H^*.

Next $[Q_e/\langle e,z\rangle, V_0] = (V_0\cap Q_e)\langle e,z\rangle/\langle e,z\rangle$ is of rank 6 and $|\tilde{Q}:\tilde{Q}_e| = 2$. Therefore $[\tilde{Q}, V_0]$ is of rank 6 or 7 if G is F_2 and of rank 6, 7, or 8 if G is F_1. However as V^* is 2-central in H^*, $V^* \le (H^*)^\infty$, so V^* is not of type b_m on $\tilde{Q}$. Thus $[\tilde{Q}, V_0]$ is of even rank. As a matter of fact if G is F_1 that rank is 8, not 6. For otherwise as $C_{H^*}(V^*)/O_2(C_{H^*}(V^*)) \cong \Omega_8^+(2)$, $C_{H^*}(V^*)$ centralizes $[\tilde{Q}, V]$ whereas this does not even hold in H_e. In particular if G is F_2 then $[Q,V] = [Q_e,V] \le V$. On the other hand if G is F_1 then we will see that $[Q,V] \le V$ and hence that $[Q,V_0] = Q\cap V \not\le V_0$ as $m([\tilde{Q},V]) = 8 = m((Q\cap V)/Z)$, and since we observed in the previous paragraph that $V = V_0(Q\cap V) >$

V_0.

Let $R_e = O_2(M_e)$. Assume in this paragraph that G is F_1. First R_eQ_e is of index 2 in R_eQ and hence normal in that group. Further $R_e/\langle e\rangle$ is special with center $V/\langle e\rangle$ and with M_e irreducible on the Frattini factor group, so $R_e/(Q\cap V)$ is extraspecial with center $V/(Q\cap V)$. Thus $V/(Q\cap V) = Z(R_eQ_e)/(Q\cap V)$ is Q-invariant. That is $[V,Q] \le V$ as claimed.

So in any event $Q \le M = N_G(V)$. Also $(H\cap M)^*$ is contained in the maximal parabolic $N_{H^*}(V^*) \cong Sp_6(2)/D^4$ or $\Omega_8^+(2)/D^4$ of H^*. But $(H\cap M)_e^* \cong \Omega_6^-(2)/D^4$ or $Sp_6(2)/D^4$ and hence is maximal in $N_{H^*}(V^*)$. Hence $H\cap M = N_H(V^*)$ or $Q(H\cap M)_e$. Next $Q\cap V$ is a hyperplane of V and $\langle Q,M_e\rangle \le M$ is irreducible on V. When G is F_1 we are using the fact that $[Q,V_0] = Q\cap V$.

Anyway by 17.6, V is an orthogonal or symplectic space and $Y/C_Y(V) = Sp(V)$ or $\Omega(V)$, respectively, where $Y = \langle Q^M\rangle$. In particular if G is F_1 then as M does not act on V_0, $Q(H\cap M)_e$ is proper in $H\cap M$, so $H\cap M = N_H(V^*)$. Then $(H\cap M)/O_2(H\cap M) \cong \Omega_8^+(2)$, so $Y/C_Y(V) = \Omega(V) \cong \Omega_{10}^+(2)$ and $Y = M$.

Similarly if G is F_2 and $H\cap M = N_H(V^*)$ then $M/C_M(V) \cong Sp_8(2)$. On the other hand if $H\cap M = Q(H\cap M)_e$ then $R = O_2(M) = C_Q(V)R_e$ with $|R:R_e| = 2$ and $M = RM_e$. Then as $V\langle e\rangle = Z(R_e)$ and $[C_Q(V),e] = Z$, we have $V\langle e\rangle \trianglelefteq M$ and then also $Z = [V\langle e\rangle,R] \trianglelefteq M$, a contradiction.

So M/R is as described in the introduction and as $H\cap M \in \mathcal{N}_H(T)$, $M \in \mathcal{N}(T)$. Notice $R = R_e$. $V\langle e\rangle = Z(R)$ and M_e is irreducible on $R/Z(R)$, so R is of class 2. Indeed if G is F_1 then also M is irreducible on $V = Z(R)$ so R is special. If G is F_2 then M is the stabilizer of an involution in the corresponding parabolic of F_1. This tells us M is indecomposable on Z(R). It also tells us R is special, except that we used that fact inductively in our proof above. R can be seen to be special directly inside of ${}^2E_6(2)/Z_2$. For example this is stated explicitly in lemma 3.3 of [36], or it can be checked from the relations in [25].

It remains to construct the remaining parabolics. Let V_i, $1 \le i < n$ be totally singular subspaces of V with $m(V_i) = i$ unless G is F_1 and $i = 4$, where take $m(V_4) = 5$. Then $V_i \le Q$. Let $G_i = N_G(V_i)$. Then indeed $G_1 = H$ and $G_2 = N_G(U)$. Let $X_i = \langle Q^{G_i}\rangle$ and $R_i = O_2(G_i)$. As $Aut_M(V_i) = GL(V_i)$, $X_i \le M$ and $G_i/R_i = (X_i/R_i) \times (C_G(V_i)/R_i)$. Let $H_i = G_i\cap H$. We need to show $\mathcal{N}_H(T) = \{H_i : 2 \le i \le n\}$, determine the structure of G_3, and if G is F_1, determine the structure of G_4.

Let $E = Q\cap Q^g$. From the structure of G_2, E/U is the Todd module for $H_2/O_2(H_2)$. In particular from section 31, V_i/U is the unique $(m(V_i)-2)$-dimensional subspace of E/U fixed by T and $C_G(V_i)/R_i$ is isomorphic to S_5 or S_6/Z_3 if $i = 3$ and G is F_2 or F_1, respectively, and to S_3 if $i = 4$. Therefore $H_i = \langle C_G(V_i), N_{H\cap M}(V_i)\rangle$ is a maximal parabolic of H, as desired.

If G is F_1 and $V_3 = \langle U, z^h\rangle$ then $V_4 \leq D = E\cap Q^h$ and $U\langle (V_4)^{C(V_3)}\rangle = D_0$ is of rank 9, and as $C(V_3) \leq N(D)$ we have $D_0 \leq D$. Also $E\cap V \not\leq D$ while $C(V_3)$ is irreducible on E/D_0. Therefore $D = D_0$. The 2-chief factors for G_3 can now be recovered from this information and 17.7. Using the structure of G_3 and the fact that $G_{3,e}/\langle e\rangle$ is the third parabolic in F_2, we obtain the strucure of that parabolic of F_2.

Next $G_4\cap C(V_3)$ is irreducible on D/V_4 and $D\cap V \not\leq Q^y$ for $z^y \in V_4 - V_3$, so V_4 is the intersection of the groups Q^y as z^y varies over V_4. On the other hand if $W = \langle V_3, z^k\rangle$ is a hyperplane of V_4 then working in M, $D\cap Q^k\cap V = B$ is of rank 6, (as $D\cap Q^k$ is the intersection of 4 M-conjugates of Q) $A = \langle B^{C(V_3)\cap N_M(W)}\rangle$ is of rank 7, and $C(V_3)\cap N_M(W)$ is irreducible on D/A. Thus $A = D\cap Q^k$ as $D\cap V \not\leq Q^k$. This information together with 17.7 says a 2-chief series for G_4 is as described in the introduction and hence R_3 is of class 3.

31. The Todd module.

Section 31 lists various facts about the smallest faithful modules in characteristic 2 for the Mathieu groups M_{24}, M_{23}, and M_{22}. Presumably all results listed already appear in the literature, but proofs are at least sketched here for completeness.

In this section G is M_{24} and Ω is a set of order 24 upon which G acts transitively. Let E be the permutation module for G on Ω over GF(2). Thus E is a vector space over GF(2) with basis $(x_i : i \in \Omega)$ permuted transitively by G via $(x_i)g = x_{ig}$ for $g \in G$. For $\Delta \subseteq \Omega$ write $e(\Delta)$ for the element $\sum_{i\in\Delta} x_i$ of E.

In particular $e(\Omega)$ is a fixed point of G and $E_0 = \{e(\Delta) : |\Delta| \text{ even}\}$ is a G-invariant hyperplane of E. Define $\bar{E} = E/\langle e(\Omega)\rangle$. There is a standard inner product (,) and quadratic form Q on E_0. Q has the property that $e(\Delta)$ is singular precisely when $|\Delta| \equiv 0 \bmod 4$. Thus the element $e(\Omega)$ is singular with respect to Q. The innner product (,) is nondegenerate on $\bar{E}_0$. (cf. Exercise 7.7 in [6] for all these remarks.)

There is a 12-dimensional G-submodule D of E called the *Galoy code*. D is generated by the vectors $(e(\Delta) : \Delta \in \mathcal{O})$, where $\mathcal{O}$ is a collection of 759 8-subsets of Ω called *octads*. (cf. [19], [43]) Set $\tilde{E} = E/D$.

It turns out that $|\Delta| \equiv 0 \bmod 4$ for all Δ with $e(\Delta) \in D$. (cf. Theorem 8 in [19].) Thus $\bar{D}$ is a maximal totally singular subspace of $\bar{E}_0$. Therefore (,) defines a GF(2)G-isomorphism between the dual space $(\bar{D})^*$ of $\bar{D}$ and $\tilde{E}_0$.

Let $U = \tilde{E}_0$ and $V = \tilde{E}$. U is an 11-dimensional irreducible and V a 12-dimensional indecomposable with $U = [V,G]$. The modules V and U are studied by Todd in [43]. Hence we shall call U the *Todd module* for G. By the last paragraph $\bar{D}$ is GF(2)G-isomorphic to the dual U^* of U, so we'll call $\bar{D} \cong U^*$ the *dual Todd module*.

(31.1) (1) $H^1(G,U) \cong Z_2$.

(2) V is the unique maximal GF(2)G-indecomposable with $[V,G] = U$.

(3) D is GF(2)G-somorphic to V^*.

Proof. As $\bar{D} \cong U^*$ and D is indecomposable (cf. Exercise 7.7 in [6]) it follows that (2) implies (3). As V is indecomposable, (1) is equivalent to (2). See section 17 in [6] for these last two remarks.

Suppose I is an indecomposable GF(2)G-module with $[I,G] = U$ and $m(I) = 13$. Let $X,Y \leq G$ be of order 23, 11 with $Y \leq N_G(X)$. Then $U = [V,X]$ and $[V,Y]$ is a hyperplane of U. Further there is an involution $t \in G$ inverting Y and from the list of maximal subgroups of G in [19], $G = \langle X,t\rangle$. However $[C_I(Y),t] \leq C_U(Y)$ and $C_U(Y)$ is of rank 1, so t centralizes a hyperplane of $C_I(Y)$. Hence $0 \neq C_I(Y)\cap C(t)\cap C(X) \leq C_I(\langle X,t\rangle) = C_I(G)$, contrary to the assumption that I is an indecomposable.

The statements in the next two lemmas can be found in [19] and [43], or alternatively can be easily derived.

(31.2) Let U be the Todd module and V its 12-dimensional indecomposable cover. Then

(1) G has four orbits v_i^G on $V^\#$ of length $\binom{24}{i}$, $1 \leq i \leq 3$, and $\binom{24}{4}/6$, respectively. Further $v_i = \tilde{e}(\Delta_i)$ for Δ_i of order i. Thus $v_2,v_4 \in U$ and $v_1,v_3 \in V - U$.

(2) The stabilizer in G of v_i is isomorphic to M_{23}, $Aut(M_{22})$, $S_3/L_3(4)$, and $S_6/Z_3/E_{64}$, respectively.

(31.3) Let $W = \bar{D}$ be the dual Todd module. Then

(1) G has four orbits d_i^G on $D^\#$ of length 1,759,2576, and 759. Indeed $d_1 = e(\Omega)$, $d_2 = e(\Delta)$ for $\Delta \in \mathcal{O}$, $d_3 = e(\alpha)$, α of order 12, and $d_4 = d_1+d_2$.

(2) The stabilizer in G of d_2, d_3, is isomorphic to A_8/E_{16}, M_{12}, respectively.

(3) The stabilizer of $\bar{d}_3$ is isomorphic to $Aut(M_{12})$.

(31.4) There are exactly two 11-dimensional irreducible GF(2)G-modules. A Sylow 2-group T of G fixes a unique point of each such module. The stabilizer of a point fixed by T is $S_6/Z_3/E_{64}$ in the Todd module and A_8/E_{16} in the dual Todd module.

Proof. By James [27] there are just two 11-dimensional irreducibles for G. The remaining remarks follow from 31.2, 31.3, and the fact that T is selfnormalizing.

Let $L = M_{23}$. An element of order 23 in L is irreducible on U and U^*, so each module is an irreducible GF(2)L-module. These modules will be called the *Todd* *module* and *dual Todd module*, for M_{23}, respectively.

Similarly $K = M_{22}$ fixes a point v_2 of U by 31.2.2. An element of K of order 11 is irreducible on $U/\langle v_2\rangle = U_2$, so U_2 is an irreducible GF(2)K-module called the *Todd module* for M_{22}. By duality, K is irreducible on a hyperplane U_2^* of U^*; U_2^* is the *dual Todd module* for M_{22}.

(31.5) Let U be the Todd module for $L = M_{23}$. Then L has three orbits u_i^L on $U^\#$. u_1 and u_2 are in v_2^G and have stabilizer M_{22} and $Z_2/L_3(4)$ in L, respectively. $v_4^G = u_3^L$ and u_3 has stabilizer $Z_2/(Z_3 \times L_2(4))/E_{16}$ in L.

Proof. This follows pretty easily from 31.2 and the action of L on Ω. Notice $O_{2,3}(L_{u_3})$ is not in the center of $L_{u_3}/O_2(L_{u_3})$ as stated in [19].

Similar arguments establish the next three lemmas.

(31.6) $L = M_{23}$ has three orbits b_i^L, $1 \leq i \leq 3$, on the points of the dual Todd module U^*. b_1 and b_2 are in $\bar{d}_2^G$ while $b_3^L = \bar{d}_3^G$. The stabilizers are A_7/E_{16}, A_8, and M_{11}, respectively.

(31.7) M_{22} has three orbits on the points of the Todd module. The stabilizers are $L_3(4)$, S_5/E_{16}, and $Z_4/E_9/E_{16}$.

(31.8) M_{22} has three orbits on the points of the dual Todd module. The stabilizers are A_6/E_{16}, $L_3(2)/E_8$, and M_{10}, respectively.

(31.9) Let U and W be the Todd and dual Todd modules for $G \cong M_{24}$. Let G_i, $1 \leq i \leq 3$, be the maximal subgroups of G containing a Sylow 2-subgroup T of G with $G_1 \cong S_6/Z_3/E_{64}$, $G_2 \cong (S_3 \times L_3(2))/E_{64}$, and $G_3 \cong A_8/E_{16}$. Then G_i is indecomposable on U and W and

(1) G_1 has chief series 4/6/1 on U and 1/6/4 on W.

(2) G_2 has chief series 2/6/3 on U and 3/6/2 on W.

(3) G_3 has chief series 1/4/6 on U and 6/4/1 on W.

Proof. As W is the dual of U it suffices to prove each result for just one of the modules.

T fixes a unique point u of U and the stabilizer of that point is G_1. Therefore G_i is irreducible on $U_i = \langle v^{G_i}\rangle$ for each i. Moreover G_{13} is the parabolic of G_3 with $G_{13}/O_2(G_{13}) \cong S_3 \times S_3$. Therefore V_3 is the 6-dimensional $\Omega_6^+(2)$ module for $G_3/O_2(G_3)$. (cf. [21]) Similarly V_2 is of dimension 3, W_1 is of dimension 4, and W_2 is of dimension 2. Also by duality, T fixes a unique hyperplane U_3' of U and W_1' of W with G_3 and G_1 the stabilizers of the respective hyperplanes. An element of order 5 in G_3 has a 3-dimensional fixed point space and hence is irreducible on U_3'/U_3. Thus G_3 has chief series 1/4/6 on U. As U_3' is the unique hyperplane of U fixed by T, G_3 is indecomposable on U/U_3. Similarly as T fixes a unique point of U, G_3 is indecomposable on U_3'. Therefore G_3 is indecomposable on U.

Similar arguments handle the remaining parts of the lemma.

(31.10) Let $X = \mathrm{Aut}(M_{22})$, M the Todd module for X, and X_1 and X_2 the maximal overgroups of a Sylow 2-subgroup T of X with $X_1 \cong S_5/E_{32}$ and $X_2 \cong S_6/E_{16}$. Then M is an indecomposable X_i-module and

(1) X_1 has chief series 4/4/1/1 on M. $Z = C_M(X_1) = C_M(O_2(X_1))$ is of rank 1. There is a unique 6-dimensional X_1-submodule M_1 with $O_2(X_1)$ inducing the group of transvections with center Z on M_1. $X_1/O(X_1)$ is indecomposable on M_1/Z and M/M_1 is the $O_4^-(2)$-module for $X_1/O_2(X_1)$.

(2) X_2 has chief series 1/4/1/4 on M. $M_2 = C_M(O_2(M_2))$ is a rank 5 indecomposable for $M_2/O_2(M_2)$, there is a unique M_2-invariant hyperplane Z' of M, and $O_2(X_2)$ induces the group of transvections with axis Z'/M_2 on M/M_2.

Proof. Recall that we may regard A as the stabilizer in $G = M_{24}$ of a point u in the Todd module U for G, and regard M as $U/\langle u\rangle$. Let $G_i \leq G$ with $G_1 \cong S_6/Z_3/E_{64}$ and $G_2 \cong A_8/E_{16}$. We may take $X_i = X \cap G_i$ with u contained in the 7-dimensional G_1-sumodule of U and in the 6-dimensional G_3-submodule of U. Lemma 31.10 is now easily derived from 31.9.

(31.11) Let X be M_{24} or $\mathrm{Aut}(M_{22})$ and M the Todd or dual Todd module for X. Then $m(A) < m(M/C_M(A))$ for each nontrivial elementary abelian 2-subgroup A of X.

Proof. Let $Y \leq X$ with Y isomorphic to A_8/E_{16}, $S_6/Z_3/E_{64}$, S_6/E_{16}, or S_5/E_{32}, for $X \cong$

M_{24} and M Todd or dual Todd, or $X \cong Aut(M_{22})$ and M Todd or dual Todd, respectively in that order. Let $B = O_2(Y)$ and $W = C_M(B)$. From the previous two lemmas, W is of rank 6,4,5,4, respectively. Further Y contains a Sylow 2-subgroup of X so if the lemma is false then there is $E_{2^n} \cong A \leq Y$ with $n \geq m = m(M/C_M(A))$. But from the action of Y on W, $m(A/C_A(W)) \leq m(W/C_W(A))$, so setting $D = A \cap B$ we have $m(D) \geq m(M/C_M(D))$. Thus we may assume $A \leq B$.

We observe next that if t is a 2-central involution in X then $m([M,t]) = 4$ while if t is a non-2-central involution then either $m([M,t]) = 5$ or $t \notin E(X)$, $C_X(t) \cong L_3(2)/E_{16}$, and $m([M,t]) = 3$. In particular as $m \geq m([M,a])$ for $a \in A^{\#}$, it follows that $n \geq 4$ and if A contains an element a centralizing an element of order 5, then $n \geq 5$. We conclude that either $A = B$ or M is the dual Todd module, and in the latter case if X is $Aut(M_{22})$ then $A = E(X) \cap B$. In particular either $n = m([M,a])$ for some $a \in A^{\#}$ or X is M_{24}, M is the dual Todd module, and $A = B$. In the first case it is easy to see from the action of Y on M described in 31.9 and 31.10 that $C_M(A) \neq C_M(a)$, a contradiction. In the second $W = C_M(A)$ so $n < m$, again a contradiction.

32. Geometries for $^2E_6(2)$ and M(22).

In this section let G be $^2E_6(2)$, $T \in Syl_3(G)$, $E = J(T)$, Q the large symplectic subgroup of G in T, and $T \leq A \leq K \leq L \leq G$ with $L \cong M(22)$, $K \cong P\Omega_7(3)$, and $A = N_G(E)$ $\cong Z_2/\Omega_5(3)/E_{3^5}$. For $X \leq L$ let $H_X = H \cap X$. Let $\Gamma = \Gamma(G,\mathcal{N}_G(T))$ be the 3-parabolic geometry of $^2E_6(2)$ and $\Gamma' = \Gamma(L,\mathcal{N}_L(T))$ the 3-parabolic geometry of M(22). Let $\Gamma^* = \Gamma(K,\{A,N_K(Q)\})$. Notice Γ^* is the geometry of singular points and lines in 7-dimensional orthogonal space over GF(3). I'll identify each coset Xg in each geometry with its stabilizer X^g. Define the *lines* of the respective geometry to be the normalizer of Q in the respective group. Objects in each geometry will be said to be *collinear* if they are incident with a common line of the geometry.

(32.1) (1) If $Q \leq X \leq G$ with $X \not\leq N_G(Q)$ and $T \cap X \in Syl_3(X)$, then $T \leq X$.

(2) If $Q \leq Y \leq G$ with $|H:H_Y| \equiv 0 \bmod 3$ then $Y \leq H$.

Proof. By 26.5, $Q^g \leq T \cap X$ for $Z(Q)^g \in \mathcal{Y}(T)$. But by 29.6, $|(Q^g \cap H)Q| = |T|$, so the lemma holds.

(32.2) (1) L is rank 3 on the cosets of K with nontrivial orbits $\Delta = \Delta(K)$ and $\Sigma = \Sigma(K)$ of length 10920 and 3159, respectively.

(2) Δ consists of those conjugates of K collinear with K and distinct from K.

(3) For $J \in \Delta$, $K \cap J$ is of index 12 in H_L.

(4) For $J \in \Sigma$, $K \cap J \cong Sp_6(2)$.

(5) A has two orbits Σ_1 and Σ_2 on Σ of length 3^5 and $3^5.12$, respectively.

(6) For $J \in \Sigma_1$, $N_A(J)$ is a complement to E in A.

(7) For $J \in \Sigma_2$, $N_E(J) \cong Z_3$ and $N_A(J) = N_A(N_E(J)) \cong S_3 \times S_6$.

(8) Σ_1 consists of those $J \in \Sigma$ incident to $K' \in \Gamma$ with $K' \cap K = A$.

Proof. The first remark is due to Fischer. I include a proof here for completeness. First H_L acts as S_4 on the cosets of H_K; cf. the discussion of $^2E_6(2)$ and M(22) in section 23. Therefore for $g \in H_L - K$, $H \cap K \cap K^g$ is of index 12 in H_L. Thus by 32.1, $K \cap K^g \leq H$. Hence $K \cap K^g$ is of index 12 in H_L, and therefore $\Delta = (K^g)^K$ is of length 10920. By construction the members of Δ are collinear with K. As H_L acts as S_4 on H_L/H_K and K is transitive on lines through K, K is transitive on objects of type K collinear with K. Thus (2) holds.

Next as $|K^L|$ is even, a Sylow 2-subgroup S of K fixes a conjugate K^x of K distinct from K. Now

$$|K:K \cap K^x| \leq |K^L| - 1 - |\Delta| = 3159 = |K:K_0|,$$

where $K_0 \cong Sp_6(2)$. But from [15], $|K:K_0|$ is minimal subject to $S \leq K_0 < K$, so (1) and (4) are established.

By 23.1, there is a unique $K' \in \mathcal{N}_L(T) - \{K\}$ with $K' \cong P\Omega_7(3)$. Moreover $A = K \cap K'$. Recalling that Γ^* is the geometry of points and lines of 7-dimensional orthogonal space over GF(3), we see that K' is rank 3 on the cosets of A and that if $x \in K'$ with A^x not collinear with A in $\Gamma(K', \{A, H_{K'}\}) \cong \Gamma^*$, then the stabilizer in A of A^x is a complement B to E in A. Observe that if A^y is collinear to A then K^y is collinear with K. On the other hand 5 divides $|B|$ but not $|H|$, so by (2) and (3), K^x is not collinear with K. Thus if we let $\Sigma_1 = (K^x)^A$ we see that (6) and (8) hold, and that Σ_1 is of order 3^5. Hence if we define $\Sigma_2 = \Sigma - \Sigma_1$, certainly Σ_2 is of order $3^5.12$.

Let Y be of order 5 in B. Observe Y fixes 4 points of Σ because $4 = |N_K(Y):K \cap K^x \cap N(Y)|$. Further $I = N_E(Y)$ is of order 3 so Y fixes $|I| = 3$ points of Σ_1. Hence Y fixes a unique point K^t of Σ_2 and I also fixes K^t. Indeed Y is irreducible on $[E,Y]$, so $N_E(K^t) = E$ or I, and as $m_3(Sp_6(2)) = 3$ it must be the latter. Therefore $N_A(K^t) \leq N_A(I)$ so $N_A(K^t)E/E \leq Z_2 \times S_6$. On the other hand $|A|/|S_3 \times S_6| = |\Sigma_2|$, completing the proof of the lemma.

(32.3) In each geometry:

(1) Distinct objects α and β of the same type are incident with at most one common line $\alpha+\beta$.

(2) If γ is an object of the geometry incident with distinct collinear objects α and β of the same type then γ is incident with $\alpha+\beta$.

Proof. This is obvious in Γ^* as Γ^* is the geometry of singular points and lines of an orthogonal space. So consider $\theta = \Gamma(X,\mathcal{N}_X(T))$, for $X = G$ or L. We may assume α and β are distinct objects of θ and H_X is a line incident with α and β. As in the proof of 32.2, $Q \le X_{\alpha\beta}$ and $X_{\alpha\beta} \cap H$ is of index 12 in H_X. Therefore 32.1.2 implies (1).

Assume the hypothesis of (2) with $\alpha+\beta = H$. Then by (1), $G_{\gamma\alpha\beta} \le H$, so $G_{\psi\alpha\beta}$ is solvable. But if γ is not incident to H_X then in the geometry of G_γ, α and β are not collinear, so $G_{\gamma\alpha\beta} \cong Sp_6(2)$ or $P\Omega_5(3)$ for $X = G$ or L, respectively. Of course this contradicts the solvability of $G_{\gamma\alpha\beta}$ and establishes (2).

(32.4) G is flag transitive on Γ.

Proof. If not there exists a flag F which is not conjugate to a subflag of $\mathcal{N}_G(T) = \{L,L',L^*,H\}$. Then F is of rank at least 3 and by 27.6, F is not in the residue Γ_U for any $U \in F$. In particular F contains no line, since every set of objects of distinct types is a flag in Γ_U if U is a line.

So we may take $F = \{L,L',\alpha\}$, with $\alpha = L^*g$ for some $g \in G - (L \cap L')$. By the last paragraph, α and L' are not collinear in Γ_L. On the other hand α is incident in Γ_L with some L-conjugate J of L'.

Claim L' and Y are not collinear for any $Y \in J^L$ which is incident with α in Γ_L; in particular L' and J are not collinear. For if so by 32.3, $L'+Y$ is incident to L and α, contrary to an earlier remark. Let Σ_2 consist of those L-conjugates of J not collinear with J and not incident in Γ_L with α. By 32.2, $(L \cap J)_\alpha$ is transitive on Σ_2. Thus for each $X \in \Sigma_2$, no member Y of the set $\Delta(X)$ of objects in J^L collinear with X in Γ_L is collinear with α in Γ_L, since we just established this fact for $X = L'$. In particular $\Delta(X) \cap \Sigma(J) \subseteq \Sigma_2$ for each $X \in \Sigma_2$. Thus $\Sigma(J)$ is disconnected as a subgraph of the graph on J^L obtained by joining J^Y to the members of $\Delta(J^Y)$. But this is not the case by Theorem 4 of [1].

33. A 112-dimensional $GF(2)J_4$-module.

The point of this section is to show J_4 has a $U_3(11)$-subgroup, assuming the existence of a faithful $GF(2)J_4$-module of dimension 112. The existence of such a subgroup may already be known via the Norton et al construction of J_4, but I haven't found the result stated in the literature. In any case an independent proof is probably useful.

(33.1) Let $L \cong L_2(11)$, $T \in Syl_2(L)$, $y \in T^{\#}$, and x of order 3 in $N_L(T)$. Let ω be a generator of GF(4). Then

(1) L has two quasiequivalent 5-dimensional irreducible modules V and V' with field of definition GF(4) and a 10-dimensional module $\hat{V}$ with field of definition GF(2). All other irreducible GF(2)L-modules are of dimension at least 12.

(2) $\dim([V,y]) = 2$, $\dim([\hat{V},x]) = 6$, and $\dim([\hat{V},y]) = 4$. ω and ω^{-1} are eigenvalues of x of multiplicity 2 on V.

(3) $\dim(C_V(T)) = 2$ and x has eigenvalues ω and ω^{-1} on $C_V(T)$.

Proof. See James [27], page 78, for (1). [27] also contains the characters of V and $\hat{V}$, from which (2) can be derived. We conclude from (2) that if z is of order 3 in $C_L(y)$ then $[V,y] = [V,y,z]$. Hence as z is inverted by $t \in T - \langle y\rangle$, it follows that $\dim(C_V(T)) = 2$.

Let $T\langle x\rangle \le K \le L$ with $K \cong A_5$. As $\dim([V,x]) = 4$, K has two 2-dimensional chief factors and one 1-dimensional chief factor. If K fixes a point of V then V is a homomorphic image of the 11-dimensional permutation module. But from [27], $\hat{V}$ is the unique noncentral chief factor of that permutation module. So an irreducible K-submodule I of V is of dimension 2. In particular, replacing ω by ω^{-1} if necessary, x has eigenvalue ω on $C_I(T)$. Let $V^* = V/I$. $\dim(C_{V^*}(T)) = 2$ so $C_{V^*}(K) \neq 0$. As K fixes no point, $C_{V^*}(T)/I \neq C_V(T)^*$, so $[C_V(T),x] = C_V(T)$. If x has just one eigenvalue on $C_V(T)$ then $T\langle x\rangle$ fixes five points of V, each in a different L-orbit. However an easy counting argument shows there are only two orbits with a representative stabilized by $T\langle x\rangle$. This establishes (3).

(33.2) $H^1(L,\hat{V}) = 0$.

Proof. Let $U = \hat{V}$. If $H^1(L,U) \neq 0$ there is an 11-dimensional indecomposable I for L with U = [U,L]. (cf. section 17 in [7].) Let A and B be of order 11 and 5 in L, respectively, with $AB = N_L(A)$. Then U = [A,I] and $J = C_I(A)$ is a point. Similarly $C_I(B) = J + C_U(B)$ with $\dim(C_U(B)) = 2$. Finally there is an involution y inverting B, and I'll show $[C_I(B),y] = 0$. Hence L = <AB,y> centralizes J, contradicting I indecomposable.

Now y centralizes x of order 3. By 33.1.2, $\dim(C_I(x)) = 5$ so $\dim(C_I(xy)) \geq 3$. On the other hand by 33.1.2, $\dim(C_U(y)) = 6$. As $\dim([C_U(y),x]) \geq \dim([U,x])/2 = 3$ and $\dim([C_U(y),x])$ is even, we have $\dim([C_U(y),x]) \geq 4$. Therefore $\dim([C_U(y),x]) = 4$, $\dim(C_I(xy)) = 3$, and $C_I(y) = 7$.

Finally $\dim([U,B,y]) = \dim([U,B])/2 = 4$, so as $\dim(C_I(y)) = 7$, y centralizes $C_I(B)$ as desired.

(33.3) Let U be an 11-dimensional vector space over $F = GF(2^{10})$, $11^{1+2} \cong Q \leq GL(U)$, and $L \cong SL_2(11)$ a complement to Q in $N_{GL(U)}(Q)^{\infty}$. Let <t> = Z(L), x and y of order 3 and 4 in L, respectively, and z of order 3 in Z(GL(U)). Then

(1) $\dim(C_U(t)) = 6$.

(2) [U,t] is FL-isomorphic to $F\otimes V$ and L is indecomposable on $C_U(t)$.

(3) $C_U(y) \leq [U,t]$ so $(C_U(t)/W)\cap C(y) \leq [U,t]/W$ for each $W \leq [U,t,y]$.

(4) $\dim(C_U(x)) = 3$.

(5) Up to conjugation there is a unique Q_8-subgroup T of $N_{GL(U)}(Q)$. Moreover $T \in Syl_2(L)$.

(6) Up to conjugation there are three $SL_2(3)$-subgroups of $N_{GL(U)}(Q)$. They are T<x>, T<xz>, and $T\langle xz^{-1}\rangle$. $\dim([U,t,x]) = 4$ and $\dim([U,t,xz]) = 3$.

Proof. $U = \bigoplus_{D\in\Delta} C_U(D)$, where Δ is the set of noncentral subgroups of order 11 in a subgroup P of index 11 in Q. Further $C_U(D)$ is a point for each $D \in \Delta$. Now t inverts a unique $E \in \Delta$ and t has 5 orbits of length 2 on $\Delta - \{E\}$, from which (1) follows. Then $L^* = L/\langle t\rangle$ is faithful on [U,t] and $U/C_U(t)$ so the first remark in (2) follows from 33.1.1.

If L is not indecomposable on $C_U(t)$ then L fixes a point W in $C_U(t)$. But there is A of order 11 in $C_L(P)$, so A acts on $C_U(D)$ for each $D \in \Delta$ and as $[U,t] = [U,t,A]$, $W = C_U(D)$ for some $D \in \Delta$. This is impossible as $N_L(A) = N_L(P) \geq N_L(C_U(D))$.

So (2) is established. Its easy to check that if I is a 6-dimensional indecomposable with $[I,L] = V$ then $C_I(y) \leq V$, from which (3) follows. Part (4) follows from (2) and 33.1.2. Parts (5) and (6) are straightforward; use 33.1.2 for the last part of (6).

(33.4) Let G be J_4 and M a 112-dimensional irreducible GF(2)G-module. Let $Q \in Syl_{11}(G)$, $Z = Z(Q)$, $T \in Syl_2(C_G(Z))$, X of order 3 in $N_G(T) \cap C_G(Z)$, and $\langle t \rangle = Z(T)$. Then

(1) t is 2-central in G and $C_G(t) = H \cong Aut(M_{22})/Z_3/D^6$.

(2) Let $D = O_2(H)$. Then $T = C_D(Z) \cong Q_8$, $XD = O_{2,3}(H)$, $\langle t \rangle = C_D(X)$, and $L = C_G(T)^{\infty} \cong L_2(11)/Q^5$.

(3) $C_M(Z) = C_M(Q)$ is of dimension 2 and $[C_M(Z),T] = 0$.

(4) $m(C_M(t)) = 62$.

(5) $m(C_M(D)) = 10$ and $C_M(D) = C_M(DX)$ is $GF(2)(L/O_2(L))$-isomorphic to $\hat{V}$.

(6) $m(C_{M/C_M(D)}(D)) = 30$ and $[M,t,X]C_M(D)/C_M(D) = C_{M/C_M(D)}(D)$.

(7) If $v \in C_M(Q)^{\#}$ then $C_G(v) \cong Z_2/U_3(11)$.

Proof. If t is not 2-central then $C_G(t) \cong Aut(M_{22})/E_{2^{11}}$. In particular $T = O_2(C_G(\langle t,Z \rangle)) \leq O_2(C_G(t))$, which is impossible as $O_2(C_G(t))$ is abelian while $T \cong Q_8$. Therefore (1) holds. Now (2) follows from 30.2.2.

A minimal faithful GF(2)G-module is of dimension 110, so $C_M(Z) = C_M(Q)$ is of dimension 2 and Q is irreducible on $[M,Z] = [M,Q]$ of dimension 110. As A_4 is not a subgroup of $GL_2(2)$, $[C_M(Z),T] = 0$.

As $End_{C_G(Z)}([M,Z]) = GF(2^{10}) = F$, the representation π of $C_G(Z)$ on $[M,Z]$ can be regarded as a map from $C_G(Z)$ into $A = GL_{11}(2^{10})$. In particular (4) holds by (3) and 33.3.1. Let $X = \langle x \rangle$, x_0 of order 3 in $C_A(Z\langle t \rangle)^{\infty}$, and z_0 of order 3 in $Z(A)$. By 33.3.4 and 33.3.6 either $x = x_0$ and $m(C_{[M,Z]}(x)) = 30$ or $x = x_0 z_0$ and $m(C_{[M,Z]}(x)) = 40$. Also $[M,t] = [M,Z,t]$ is isomorphic to $V^F = F \otimes V$ as an FTZ-module and $C_{[M,Z]}(y) \leq [M,t]$ and $(C_{[M,Z]}(t)/W) \cap C(y) \leq [M,t]/W$ for each $y \in T - \langle t \rangle$ and $W \leq [M,t,y]$ by 33.3.3.

Let $S \in Syl_3(C_H(X))$. Then $S \cong 3^{1+2}$ so for each faithful GF(2)S-module W, $m([W,X]) \equiv 0 \bmod 3$. But by 33.3.6, $m([M,t,X]) = 40$ or 30 for $m(C_{[M,Z]}(X)) = 30$ or 40, respectively, so we conclude $m([M,t,X]) = 30$ and $x = x_0z_0$. In particular $10 = m([C_{[M,Z]}(t)/[M,t],X])$ so as $m(C_M(t)/[M,t]) = 12$, we can apply the first remark of this paragraph to $C_M(t)/[M,t]$ to conclude $[C_M(t)/[M,t],x] = C_M(t)/[M,t]$. It follows that $C_H(X)$ is irreducible on $C_M(t)/[M,t]$. Also as 30 is the minimal dimension of a faithful $C_H(X)/\langle t\rangle$-module, we conclude $C_H(X)$ is irreducible on $[M,t,X]$.

As $[M,t]$ is FTZ-isomorphic to V^F, $m(C_{[M,t]}(T)) = 20$ by 33.1.3. As $C_M(Q) \leq C_M(T)$ and $C_{[M,Z]}(T) \leq [M,t]$ by 33.3.3, we conclude $m(C_M(T)) = 22$. By 33.1.3, x_0 has eigenvalues ω and ω^{-1} on $C_{[M,t]}(T)$. Therefore $x = x_0z_0$ has eigenvalues $\omega,1$. Hence $m(C_M(TX)) = 10$ and Z is irreducible on $C_M(TX)$.

Next as $C_H(X)$ is irreducible on $[M,t,X]$, either $[C_{[M,t]}(D),X] = 0$ or $[M,t,X] \leq C_{[M,t]}(D)$. The latter is impossible as $\dim(C_{[M,t]}(T)) < \dim([M,t,X])$. Thus $C_M(D) = C_M(TX)$ is of rank 10 as Z is irreducible on $C_M(TX)$. As $C_{[M,Z]}(T) \leq [M,t]$ and H is irreducible on $C_M(t)/[M,t]$, we have $C_M(D) = C_{[M,t]}(D)$.

Let $L = C_G(T)^\infty$. Then $Z \leq L$. We will show there is $K \leq L$ with $L = K\Omega_2(L)$ and with K centralizing a point of $C_M(Q)$. Then by Theorem B, $\langle K,Q\rangle \cong U_3(11)$, establishing (7).

Let W be a hyperplane of $C_M(T)$ containing $C_{[M,t]}(T)$. $L = L^\infty$ centralizes $C_M(T)/C_{[M,t]}(T)$, so LT acts on W. $C_M(D)$ is of rank 10 and $C_H(X)/X\langle t\rangle$ is faithful on $C_M(D)$ as $[C_M(D),Z] = C_M(D)$. So $E_9 \cong S/X$ is faithful on $C_M(D)$ and thus $m([C_M(D),s]) \equiv 0 \bmod 3$ for $s \in S - X$. In particular this holds for an element s of order 3 in L, so by 33.1.1 and 33.1.2, $C_M(D)$ is $GF(2)L/O_2(L)$-isomorphic to $\hat{V}$. Thus by 33.2, if $K \leq L$ with $L = KO_2(L)$ and $[K,w] \leq C_M(D)$ for some $w \in W - [M,t]$, then we are done.

Let $M^* = M/C_M(D)$. It remains to show that for $w \in W - [M,t]$, $L = C_L(w^*)O_2(L)$. Suppose we show $[D,W^*] \neq 0$; then as Z is irreducible on $O_2(L)/\langle t\rangle \cong E_{2^{10}}$, $O_2(L)/\langle t\rangle$ acts as the full group of transvections on W^* with axis $C_{[M,t]}(T)^*$, and hence is transitive on the points of W^* not in $[M,t]^*$, completing the proof.

Thus it remains to show $[D,W]^* \neq 0$. But if $[D,W]^* = 0$ then as H is irreducible on $C_M(t)/[M,t]$, we have $(C_M(t)^* \cap C(D))[M,t]^* = C_M(t)^*$. However there is $y \in T - \langle t\rangle$, and $C_{[M,Z]^*}(y) \leq [M,t]^*$ by an earlier remark. Thus as $[M,t]^* < C_M(t)^*$, the proof is complete.

REFERENCES

1. Aschbacher, M., A homomorphism theorem for finite graphs, Proc. AMS, 54(1976), 468-470.

2. Aschbacher, M., On the maximal subgroups of the finite classical groups, Invent. Math., 76(1984), 469-514.

3. Aschbacher, M., On finite groups of Lie type and odd characteristic, J. Alg., 66(1980), 400-424.

4. Aschbacher, M., GF(2)-representations of finite groups, American J. Math., 104(1982), 683-771.

5. Aschbacher, M., Tightly embedded subgroups of finite groups, J. Alg., 42(1976), 85-101.

6. Aschbacher, M., Finite Group Theory, Cambridge University Press, London, 1986.

7. Aschbacher, M., Flag structures on Tits geometries, Geom. Ded., 14(1983), 21-32.

8. Aschbacher, M., Finite groups generated by odd transpositions, I, II, III, IV, Math. Z.,J. Alg., 127, 26(1972), (1973), 45-56, 451-491.

9. Aschbacher, M., Weak closure in finite groups of even characteristic, J. Alg., 70(1981), 561-627.

10. Aschbacher, M., A characterization of Chevalley groups over fields of odd order, Ann. Math., 106(1977), 353-468.

11. Aschbacher, M. and Hall, M., Groups generated by a class of elements of order 3, J. Alg., 24(1973), 591-612.

12. Aschbacher, M. and Seitz, G., Involutions in Chevalley groups over fields of even order, Nagoya Math. J., 63(1976), 1-91.

13. Aschbacher, M. and Seitz, G., On groups with a standard component of known type, Osaka J. Math., 13(1976), 439-482.

14. Aschbacher, M. and Seitz, G., On groups with a standard component of known type, II, Osaka J. Math., 18(1981), 703-723.

15. Aschbacher, M., and Smith, S., Tits geometries over GF(2) defined by groups over GF(3), Comm. Alg., 11(1983), 1675-1684.

16. Buekenhout, F. and Shult, E., Foundations of polar geometry, Geom. Ded., 3(1976), 155-170.

17. Burgoyne, N., Griess, R., and Lyons, R., Maximal subgroups and automorphisms of Chevalley groups, Pacific J. Math., 71(1977), 365-403.

18. Conway, J., A quaternionic construction for the Rudvalis group, in Topics in Group Theory and Computation, Academic Press, 1977.

19. Conway, J., Three lectures on exceptional groups, in Finite Simple Groups, Academic Press, London, 1971.

20. Conway, J. and Wales, D., Construction of the Rudvalis group of order 145,926,144,000, J. Alg., 27(1973), 538-548.

21. Curtis, C., Modular representations of finite groups with split (B,N)-pairs, in Seminar on Algebraic groups and related finite groups, Springer-Verlag, Berlin, 1970.

22. Fischer, B., Finite groups generated by 3-transpositions, University of Warwick preprint.

23. Gorenstein, D. and Lyons, R., The local structure of finite groups of characteristic 2 type, Memoirs AMS, 276(1983), 1-731.

24. Griess, R., Automorphisms of extraspecial groups and nonvanishing degree 2 cohomology, Pacific J. Math., 48(1973), 403-422.

25. Griess, R., Schur multipliers of finite simple groups of Lie type, Trans. AMS, 183(1973), 355-421.

26. Hall, M. and Wales, D., The simple group of order 604,800, J. Alg., 9(1968), 417-450.

27. James, G., The modular characters of the Mathieu groups, J. Alg., 27(1973), 57-111.

28. McLaughlin, J., Some groups generated by transvections, Arch. Math., 18(1967), 364-368.

29. McLaughlin, J., Some subgroups of $SL(n,F_2)$, Illinois J. Math., 13(1969), 108-115.

30. Mitchell, H., The subgroups of the quaternary abelian linear group, Trans. AMS, 15(1914), 379-396.

31. O'Nan, M., Some evidence for the existence of a new simple group, Proc. London Math. Soc., 32(1976), 421-479.

32. Parrott, D., On Thompson's simple group, J. Alg., 46(1977), 389-404.

33. Parrott, D., A characterization of the Rudvalis simple group, Proc. London Math. Soc., 32(1976), 25-51.

34. Seitz, G., Generation of finite groups of Lie type, Trans. AMS, 271(1982), 351-407.

35. Seitz, G., The root subgroups for maximal tori in finite groups of Lie type, Pacific J. Math., 106(1983), 153-243.

36. Stroth, G., A characterization of Fischer's sporadic simple group of order $2^{41}.3^{13}.5^{6}.7^{2}.11.13.17.19.23.31.47$, J. Alg., 40(1976), 499-531.

37. Timmesfeld, F., Finite groups in which the generalized Fitting group of the centralizer of some involution is extraspecial, Ann. Math., 107(1978), 297-369.

38. Tits, J., A local approach to buildings, in The Geometric Vein, Springer, 1981, 517-547.

39. Harada, K., On the simple group F of order $2^{14}.3^{6}.5^{6}.7.11.19$, in Proceedings of the Conference on Finite Groups, Academic Press, New York, 1976.

40. Aschbacher, M., The Tits group as a standard subgroup, Math. Z., 181(1982), 229-252.

41. Seitz, G., Flag transitive subgroups of Chevalley groups, Ann. Math., 97(1974), 27-56.

42. Lyons, R., Evidence for a new finite simple group, J. Alg., 20(1972), 540-569.

43. Todd, J., A representation of the Mathieu group M_{24} as a collineation group, Ann. di Math., 71(1966), 199-238.

44. Springer, T. and Steinberg, R., Conjugacy classes, in Seminar on Algebraic groups and related finite groups, Springer, Berlin, 1970.

45. Buekenhout, F., Diagram geometries and groups, J. Comb. Theory Ser. A, 27(1979), 121-151.

46. Ronan, M., and Smith, S., 2-local geometries for some sporadic groups, Proc. Sym. Pure Math., 37(1980), 283-289.

47. Ho, C., On the quadratic pairs, J. Alg., 43(1976), 338-358.

48. Smith, S., Irreducible modules and parabolic subgroups, J. Alg., 75(1982), 286-289.

California Institute of Technology

Pasadena, Calif. 91125.

General instructions to authors for PREPARING REPRODUCTION COPY FOR MEMOIRS

For more detailed instructions send for AMS booklet, "A Guide for Authors of Memoirs." Write to Editorial Offices, American Mathematical Society, P. O. Box 6248, Providence, R. I. 02940.

MEMOIRS are printed by photo-offset from camera copy fully prepared by the author. This means that, except for a reduction in size of 20 to 30%, the finished book will look exactly like the copy submitted. Thus the author will want to use a good quality typewriter with a new, medium-inked black ribbon, and submit clean copy on the appropriate model paper.

Model Paper, provided at no cost by the AMS, is paper marked with blue lines that confine the copy to the appropriate size. Author should specify, when ordering, whether typewriter to be used has PICA-size (10 characters to the inch) or ELITE-size type (12 characters to the inch).

Line Spacing – For best appearance, and economy, a typewriter equipped with a half-space ratchet – 12 notches to the inch – should be used. (This may be purchased and attached at small cost.) Three notches make the desired spacing, which is equivalent to 1-1/2 ordinary single spaces. Where copy has a great many subscripts and superscripts, however, double spacing should be used.

Special Characters may be filled in carefully freehand, using dense black ink, or INSTANT ("rub-on") LETTERING may be used. AMS has a sheet of several hundred most-used symbols and letters which may be purchased for $5.

Diagrams may be drawn in black ink either directly on the model sheet, or on a separate sheet and pasted with rubber cement into spaces left for them in the text. Ballpoint pen is *not* acceptable.

Page Headings (Running Heads) should be centered, in CAPITAL LETTERS (preferably), at the top of the page – just above the blue line and touching it.

LEFT-hand, EVEN-numbered pages should be headed with the AUTHOR'S NAME;
RIGHT-hand, ODD-numbered pages should be headed with the TITLE of the paper (in shortened form if necessary).
Exceptions: PAGE 1 and any other page that carries a display title require NO RUNNING HEADS.

Page Numbers should be at the top of the page, on the same line with the running heads.

LEFT-hand, EVEN numbers – flush with left margin;
RIGHT-hand, ODD numbers – flush with right margin.
Exceptions: PAGE 1 and any other page that carries a display title should have page number, centered below the text, on blue line provided.

FRONT MATTER PAGES should be numbered with Roman numerals (lower case), positioned below text in same manner as described above.

MEMOIRS FORMAT

It is suggested that the material be arranged in pages as indicated below.
Note: Starred items (*) are requirements of publication.

Front Matter (first pages in book, preceding main body of text).

Page i – *Title, *Author's name.
Page iii – Table of contents.
Page iv – *Abstract (at least 1 sentence and at most 300 words).

**1980 Mathematics Subject Classification (1985 Revision).* This classification represents the primary and secondary subjects of the paper, and the scheme can be found in Annual Subject Indexes of MATHEMATICAL REVIEWS beginning in 1984.

Key words and phrases, if desired. (A list which covers the content of the paper adequately enough to be useful for an information retrieval system.)

Page v, etc. – Preface, introduction, or any other matter not belonging in body of text.

Page 1 – Chapter Title (dropped 1 inch from top line, and centered).
Beginning of Text.
Footnotes: *Received by the editor date.
Support information – grants, credits, etc.

Last Page (at bottom) – Author's affiliation.

ABCDEFGHIJ–AMS–89876